D0252454

Quantum Profiles

✣

Quantum Profiles

✣

JEREMY BERNSTEIN

PRINCETON UNIVERSITY PRESS

PRINCETON, NEW JERSEY

Copyright © 1991 by Princeton University Press
Published by Princeton University Press, 41 William Street,
Princeton, New Jersey 08540
In the United Kingdom: Princeton University Press, Oxford
All Rights Reserved

"Besso" originally appeared in the *New Yorker*,
in slightly different form, as A CRITIC AT LARGE.

Library of Congress Cataloging-in-Publication Data

Bernstein, Jeremy, 1929–
Quantum profiles / Jeremy Bernstein
p. cm.
Includes bibliographical references and index.
ISBN 0-691-08725-3 (alk. paper)
1. Quantum Theory. 2. Physicists—Interviews.
3. Physicists—Biography. I. Title.
QC174.12.B464 1991
530.1'2'0922—dc20 90-40843

This book has been composed in Adobe Laser Janson

Princeton University Press books are
printed on acid-free paper, and meet the guidelines
for permanence and durability of the Committee
on Production Guidelines for Book Longevity
of the Council on Library Resources

Printed in the United States of America by
Princeton University Press, Princeton, New Jersey

1 3 5 7 9 10 8 6 4 2

✣ *Contents* ✣

Preface vii

John Stewart Bell: Quantum Engineer 3

Epilogue 90

John Wheeler: Retarded Learner 93

Epilogue 135

Besso 143

Select Bibliography 167

Index 169

✣ *Preface* ✣

SOME YEARS AGO I had a number of conversations with the late I. I. Rabi about his life and work. Toward the end of our talks, Rabi expressed his unhappiness at the fact that the revolution in physics—above all, the quantum revolution—appeared to be having such a small impact on our general cultural life. As he put it, "It's a great pity that the general public has very little inkling of the tremendous excitement—intellectual and emotional excitement—that goes on in the advanced fields of physics. "Neither Rabi nor I had any idea how rapidly this would change. We are now awash in public excitement about the "advanced fields of physics" and, above all, about quantum mechanics. One sometimes has the feeling that much of the general public, ranging from New Age gurus to playwrights, from literary critics to futuristic economists, and on and on, have decided, as the phrase goes, that quantum mechanics is much too serious to be left to the physicists. As a physicist, I look at this curious development with a mixture of enthusiasm, amusement, and dismay. The dismay comes about because the theory is being asked questions that neither it, nor any other scientific theory, is designed to answer. Quantum theory is not New Age mysticism, it is hard science. Bell's theorem is not a mantra, it is a theorem. The Einstein-Podolsky-Rosen experiment is not done in ashrams, it is done in physics laboratories.

When, as a writer, my level of dismay is high enough I do something about it. I write. But how? For me, the most congenial way of communicating science has been through the medium of the profile. When I wanted to learn something, both for my own sake and for that of a potential reader, I would pick out one or more individuals who were identified with that field of work and go and talk to them—sometimes for hours, sometimes for days, and sometimes for weeks. Or I would read about them in such detail that, by the end, I *was* almost talking to them. That is what this book is meant to feel like—a series of conversations carried on, on the readers' behalf and my own, about the deep issues of quantum mechanics. Hence the odd title *Quantum Profiles*. After reading this book I hope that the reader will have a real sense of what these issues are and also what they are *not*. I hope that these profiles convey the "intellectual and emotional excitement" of this enterprise, but do so honestly, without any compromise in scientific integrity. I also hope they are fun to read. They gave me much joy to write.

While I cannot personally thank Einstein and Besso, the subjects of the last of these profiles, I can thank John Wheeler and John Bell, the subjects of the first two. Both of these remarkable men spent hours talking to me, and sometimes, as I was writing these profiles, I could almost hear them talking to each other. I would also like to thank several colleagues who made comments on the writing at various stages. They include L. Brown, S. Coleman, G. Feinberg, M. Gell-Mann, and A. Pais. I have learned much from them as well.

New York
February 1990

Quantum Profiles

John Stewart Bell

QUANTUM ENGINEER

In 1902 the Olympia Academy was founded in Bern, Switzerland. It had three members: Maurice Solovine, Conrad Habicht, and Albert Einstein. Solovine, a young student, had answered a newspaper advertisement offering private tutoring in physics by Einstein for three Swiss francs an hour, while Habicht, who was studying mathematics with the idea of becoming a secondary school teacher, already knew Einstein. The three young men had regular evening "Academy" meetings over the next three years, at which they studied philosophy and discussed physics. Solovine, who was Romanian, recalled having eaten caviar in his parents' home in Romania, and on one of Einstein's birthdays he and Habicht treated Einstein to some expensive caviar, which he had never tasted. As luck would have it, that was the evening that Einstein was scheduled to talk to the "Academy" about Galileo's principle of inertia. He became so absorbed that he ate all the caviar without realizing what he was eating.

In 1905 the group broke up when Habicht and Solovine left Bern. However, they began corresponding almost at once, and in one of his earliest letters to Habicht, written in the spring of 1905, Einstein described his program of research for the year in the sort of cheerful tone one might use to describe a little light reading. It was, however, this research program that laid the foundations of twentieth-century physics. He writes, "I promise you four papers . . . the first . . . deals with radiation and energy characteristics of light and is very revolutionary . . . The second work is a determination of the true size of the atom from the diffusion and viscosity of dilute solutions of neutral substances. The third proves that assuming the molecular theory of heat, bodies whose dimensions are of the order of 1/1000 mm, and are suspended in fluids, should experience measurable disordered motion, which is produced by thermal motion. It is the motion of small inert particles that has been observed by physiologists, and called by them 'Brown's molecular motion.' The fourth paper exists in first draft and is an electrodynamics of moving bodies employing a modification of the doctrine of space and time; the purely kinematical part of this work will certainly interest you." What is striking about this list—apart from the fact that it was compiled by a then totally unknown twenty-six-year-old physicist—is that the first paper announces, it turns out, the invention of the quantum, while the last paper announces the invention of the theory of relativity, and, of the two, it is only the former that is in the young Einstein's view "revolutionary."

The theory of the light quantum, which Einstein initiated in 1905 and which is with us still, is certainly the most revolutionary development in the history of physics and arguably in the history of science. The creators of the theory, men such as Einstein, Niels Bohr, Werner Heisenberg, Erwin Schrödinger, Wolfgang Pauli, and Paul Dirac, were often struck by the apparent "absurdity"—the utterly noncommonsensical aspects—of the world depicted by the quantum theory. Long after he had done this seminal work, Heisenberg recalled that "an intensive study of all questions concerning the interpretation of quantum theory in Copenhagen finally led to a complete and, as many physicists believe, satisfactory clarification of the situation. But it was not a solution which one could easily accept. I remember discussions with Bohr which went through many hours till very late at night and ended almost in despair; and when at the end of the discussion I went alone for a walk in the neighboring park I repeated to myself again and again the question: Can nature possibly be as absurd as it seemed to us in these atomic experiments."

Unlike some of the other great intellectual revolutions of the twentieth century—in art, music, literature—until recently, at least, this one was not widely known, let alone understood, by the general public. Many physicists realized this, and a few of them tried to do something about it. For example, in 1953

Robert Oppenheimer delivered a very successful series of Reith Lectures over the BBC about subatomic physics. In one of them he characterized the epoch of the discovery of the quantum theory as follows: "It was a time of earnest correspondence and hurried conferences, of debate, criticism, and brilliant mathematical improvisation." And then he adds, with a touch of the baroque eloquence of which he was a master, "For those who participated, it was a time of creation; there was terror as well as exaltation in their new insight. It will probably not be recorded very completely as history. As history, its re-creation would call for an art as high as the story of Oedipus or the story of Cromwell, yet in a realm of action so remote from our common experience that it is unlikely to be known to any poet or any historian."

Oppenheimer died in 1967, and while there has not as yet arisen such an "art as high as the story of Oedipus," in the last few years the quantum theory has, much to the surprise of most physicists, entered into the popular culture. Fiction writers now cite, rightly or wrongly, the Heisenberg uncertainty principle as a basis for theories of literature. A recent *New York Times* book review said approvingly of a novelist that "she knows enough about Heisenberg to realize that the act of observation alters the object observed; or in literary terms, telling the story alters the story being told." Tom Stoppard's latest play, *Hapgood*, turns on the uncertainty principle; the printed text is preceded by a quote on the quantum theory of Richard Feynman. Even otherwise sober and non-science-oriented magazines such as *The Economist* have felt an obligation to alert their readers that something odd has happened in physics. In a feature article in its January 7, 1989, issue entitled "The Queerness of Quanta," *The Economist* notes that "many of this century's most familiar technologies come with an odd intellectual price on their heads. The equations of quantum mechanics explain the behaviour of sub-atomic particles in nuclear reactors and of electrons in computers and television tubes, the movement of laser light in fibre-optic cables and much else. Yet quantum mechanics itself appears absurd." Among the "absurdities" the following are offered:

"There are no such things as 'things.' Objects are ghostly, with no definite properties (such as position or mass) until they are measured. The properties exist in a twilight state of 'superposition' until then."

"All particles are waves, and waves are particles, appearing as one sort or another depending on what sort of measurement is being performed."

Then, "A particle moving between two points travels all possible paths between them simultaneously."

And, finally, "Particles that are millions of miles apart can affect each other instantaneously."

While most physicists would find these capsule descriptions of the quantum theory caricatural, there is enough truth in them to explain why people who seem to have an aversion to more conventional science are drawn to the quantum theory. The quantum theory has become the basis of the New Age outlook, with its emphasis on Eastern religions and holistic medicine. This surely would have astonished Oppenheimer, who, incidentally, studied Sanskrit so that he could read the *Upanishads* in the original. Books like Gary Zukav's *The Dancing Wu Li Masters*—quantum theory with a dash of Eastern mysticism—abound, and no ashram—at least, no Western one—can afford to be without its resident expert. Just recently, in a health food store in Greenwich Village, I came across an announcement in the *I Am News* of the Ananda Ashram in Monroe, New York, which, under the heading "Quantum Dynamics," reads, "Spiritual Purification Program: Includes meditation, fire ceremony, rebirthing, sweat lodge and Quantum Dynamics initiation for those who haven't had it; includes breathing techniques and mantra to dissolve upsets. This weekend we will work with Quantum Dynamics to dissolve past life karma all the way back to original cause."

Pace Robert Oppenheimer.

Although it is always somewhat dangerous to look for the cause of a complex sociological phenomenon in a single event, nonetheless, I believe that a cause can be made for the proposition that the present widespread interest in the quantum theory can be traced to a single paper with the nontransparent title "On the Einstein-Podolsky-Rosen Paradox," which was written in 1964 by the then thirty-four-year-old Irish physicist John Bell. It was published in the obscure journal *Physics*, which expired after a few issues. Bell's paper was, as it happens, published in its first issue. Bell, who has worked since 1960 at CERN, the gigantic elementary-particle physics laboratory near Geneva, has been known to claim that his paper involves only the use of "high school mathematics"; however, its six pages are dense with an extremely abstract set of arguments, which even professionals in the field must work hard at to understand. In fact, for several years after its publication, few if any professional physicists bothered to try.

This changed dramatically in 1969, when it was realized that "Bell's theorem" (or "Bell's inequality," as it is often called) could actually be tested in the laboratory. What was at stake in such a test was nothing less than the meaning and validity of the quantum theory. If Bell's inequality was satisfied, it would mean that all of Einstein's intuitions about the essential incompleteness of the quantum theory had been right all along. If the inequality was violated, it would mean—at least, so many physicists believe—that Bohr and Heisenberg had been right all along and that no return to classical physics was possible. By the early 1970s such experiments were actually being carried out. They still are. With a few exceptions—glitches, one thinks—all these experiments show that Einstein was wrong. It was these experimental results that caused a new generation of physicists to confront just how peculiar and counterintuitive quantum mechanics really is. It is this realization that is being reflected in the growing popular interest in the theory.

As it happens, I have known John Bell for thirty years. When his tenure had just begun at CERN, in 1960, I had begun a series of visitations to the laboratory which still continues. During that time I have talked to Bell about many things, but, as it happens, very little either about his life or about his work in quantum mechanics. John Bell's wife, Mary, is also a physicist at CERN (one of the few women at CERN with any sort of academic degree). The Bells, while charming company when one gets to know them, tend to be very private people, keeping pretty much to themselves. "Mary and I are rather unsociable," Bell once remarked to me in his lilting Irish brogue. ("Mathematics" is pronounced something like "mah-thah-mahtics" and "now" comes out sounding something like "nae." Mary Bell is Scotch and has a fine "birr." Girl sounds something like "gurrrle.") Bell has a dry wit, and one must pay attention when he speaks to see that he is not teasing. The Bells have no children and give the impression of taking great and constant pleasure in each other's company.

I am quite sure that Bell would have been most willing to discuss his ideas about the quantum theory if I had asked, but somehow I never did. I had, in fact, never tried to read Bell's 1964 article. But in the last year or so, with all the mounting interest in the subject, I decided that I had been missing out on something and that I would educate myself. This process was aided no end when Bell sent me a copy of his book—his collected papers on the quantum theory which was published in 1987 under the title *Speakable and Unspeakable in Quantum Mechanics*. It contains twenty-two essays, including his 1964 paper. Some of the essays are addressed to the educated layperson, including a celebrated one written in 1981 and bearing the unlikely title "Bertlmann's Socks and the Nature of Reality." (Bertlmann, I have been told, is a real person). I quote the opening paragraph because it is illustrative of Bell's style. It reads. "The philosopher in the street, who has not suffered a course in quantum mechanics, is quite unimpressed by Einstein-Podolsky-Rosen correlations. He can point out many examples of similar correlations in everyday life. The case of Bertlmann's socks is often cited. Dr. Bertlmann likes to wear socks of different colours. Which sock he will have on a given foot on a given day is quite unpredictable. But when you see that the first sock is pink you can be already sure that the second sock will not be pink. Observation of the first, and experience of Bertlmann gives immediate information about the second. There is no accounting for tastes, but apart from that there is no mystery here. And is not the [Einstein-Podolsky-Rosen] business just the same?" That it is not—a fact that brings us deep into the mysteries of the quantum theory—is the subject of the rest of this essay and most of the other essays in the book.

Bell has always talked over his ideas with Mary, and at the end of the preface he writes, "In the individual papers I have thanked many colleagues for their help. But here I renew very especially my warm thanks to Mary Bell. When I look through these papers again I see her everywhere."

Having studied the collection and an equally rewarding one entitled *Quantum Theory and Measurement* edited by the physicists John Wheeler and Wojciech Zurek, I felt that I finally understood enough about the subject to at least carry on a sensible dialogue with Bell. I thought I would also use the opportunity to find out something about his and Mary's lives. The odd twists of fate that determine the lives of young scientists have always fascinated me, and I could not imagine what circumstances had brought the Bells from Ireland and Scotland to Geneva.

Bell has a very busy schedule, but we finally agreed upon a week in January. For a skier, this is an especially happy time to visit CERN, which is less than a half hour from the ski runs in the Jura. I did not think there would be much chance to persuade the Bells to go skiing on their lunch hours, since I knew that both of them had given up downhill skiing in favor of cross-country some years ago. They own a modest apartment in Champéry, a ski resort not far from Geneva, where they can cross-country ski in peace. But I thought I would spend alternate lunch hours downhill skiing in the Jura and eating in one of the CERN cafeterias with the Bells. As the plane landed in Geneva I saw that the Jura was brown—hardly a trace of snow. It later turned out that there was no skiing at all.

An informal tradition has developed at the laboratory, according to which "aristocrats" like theoreticians eat lunch late—one o'clock or so—while the more "plebian" members of staff eat earlier. The Bells traditionally have lunch promptly at 11:45, and eat with the same small group consisting of, among others, a Dutch computer expert and a Norwegian in charge of laboratory safety. Both Bells are vegetarians, Mary since childhood and Bell since the age of sixteen. Since there was, at least in the past, no particular provision for vegetarian diets in the CERN cafeterias, Mary Bell would usually arrive at lunch with a satchel of fresh vegetables, which she and John would share. There is also the very pleasant tradition at CERN of after-lunch espresso in the large lounge next to the main cafeteria, or outside on the patio, from which one has a view of Mont Blanc, if the weather is nice. Gossip after lunch is another tradition at CERN.

CERN is located on the French border, a few miles from Geneva. When I first went there, it was entirely in Switzerland. As the accelerators—the giant elementary particle accelerators—that are the main business of CERN—got larger and larger, the terrain of the laboratory spilled over into France. The tunnels that contain the evacuated pipes in which the beams of elementary particles run, cross the border in several places. This has reached a kind of apotheosis in an accelerator called the LEP (Large Electron Project), whose construction was begun in 1981. By the time it was turned on in the summer of 1989, it had cost about a billion Swiss francs (nearly three quarters of a billion dollars), which was provided by the thirteen European member states that run CERN. The LEP tunnel has a radius of twenty-seven kilometers. The tunnel contains electrons and their antiparticles—positrons—streaming in opposite directions. From time to time the beams collide, and the results are the highest-energy electron and positron collisions ever observed.

Because of all this activity, the character of the laboratory has changed dramatically from what I remember of my first visit nearly thirty years ago. This was not so long after the war, and the European community had not as yet rebuilt its scientific establishment; indeed, CERN was conceived to accelerate that process. Those days are long past. The only American project that can rival the LEP is the Super Collider in Texas, if it ever gets built.

While there has been an enormous amount of construction at CERN in the last thirty years, the theoreticians still occupy part of the same compact-looking four-story building that they did when I first went there. I had no trouble finding Bell's office, the same one I had been stopping in, on and off for several decades. Like the rest of the laboratory, the theoretical division—TH, as it is called—has undergone an almost exponential expansion. When I first visited it, the entire division, including visitors like myself, consisted of something like thirty people. Now there are about a hundred and forty.

The signs on Bell's door read "J.Bell" and "M.Bell." I knocked and was invited in by Bell. He looked about the same as he had the last time I saw him, a couple of years ago. He has long, neatly combed red hair and a pointed beard, which give him a somewhat Shavian figura. On one wall of the office is a photograph of Bell with something that looks like a halo behind his head, and his expression in the photograph is mischievous. Theoretical physicists' offices run the gamut from chaotic clutter to obsessive neatness; the Bells' is somewhere in between. Bell invited me to sit down after warning me that the "visitor's chair" tilted backward at unexpected angles. When I had mastered it, and had a chance to look around, the first thing that struck me was the absence of Mary. "Mary," said Bell, with a note of some disbelief in his voice, "has *retired*." This, it turned out, had occurred not long before my visit. "She will not look at any mathematics now. I hope she comes back," he went on almost plaintively; "I need her. We are doing several problems together." In recent years, the Bells have been studying new quantum-mechanical effects that will become relevant for the generation of particle accelerators that will perhaps succeed the LEP. Bell began his career as a professional physicist by designing accelerators, and Mary has spent her entire career in accelerator design. A couple of years ago Bell, like the rest of the members of CERN theory division, was asked to list his physics speciality. Among the more "conventional" entries in the division such as "superstrings," "weak interactions," "cosmology," and the like, Bell's read "quantum engineering."

Bell thought we would be more comfortable in our discussions if we found a larger office. It turned out that the one next to his was temporarily free. Mary, he told me, would be glad to talk to me over the phone if I had something I wanted to ask her. We settled into the new office, and I asked Bell if he would mind telling me a bit about his early life and how he got into physics. "I was born on July 28, 1928, in Belfast," he began. "My parents were poor but honest. Both of them came from the large families of eight or nine that were traditional of the working-class people of Ireland at that time. Both sides of the family have been in Northern Ireland for many generations. But we are from the Protestant tribe—the British side—so the real Irish people regard us as colonists." I was curious, as I always am, as to whether there had been any scientific or academic tradition in his family. He thought a moment and replied, "As far as I know, until the present generation, there was none. The kind of professions I had heard about in the family were carpenters, blacksmiths, laborers, farm workers, and horse dealers. My father's first profession was horse dealer. He stopped going to school at the age of eight—his parents paid fines from time to time for that. He learned how to buy and sell horses instead. The nearest I heard of anyone in my family being educated was my mother's half brother. He was a village blacksmith, but he taught himself something about electricity at a time when not many people knew about electricity. I was the only one of my siblings who reached high school. I have an older sister and two younger brothers, and they left school at about the age of fourteen. The normal thing would have been for me to get a job when I reached fourteen."

Encouraged by his mother, Bell applied for financial help to go to secondary school. At the time, there was no universal system of free secondary education in Britain. That would come a bit later with the Labour government. "I sat many examinations," Bell recalled, "for the more prestigious secondary schools, hoping for scholarships, but I didn't win any." Some money did appear—Bell is not sure from where—so that he was able to attend the Belfast Technical High School, the least expensive. Bell remembers it with great affection. He did courses in bricklaying, carpentry, and bookkeeping along with the more conventional curriculum.

Unlike many prominent theoretical physicists I have asked, Bell does not have any early memories of scientific or mathematical precocity. He does, however, recall that at the age of fourteen he began a brief phase of reading Greek philosophers. "I was a bookish sort of child, much in the local public libraries. I was hostile to the idea of sports," he said laughing. "I regret it now. I've grown up to be a seven-stone weakling. I was, you know, brought up in the Church of Ireland. I was even confirmed by a bishop. But in my adolescent years I began to wonder if it was really true what they told you. Does God exist? Questions like that. So, like many children, I started looking for answers. One place to look, I thought, was philosophy. I was reading thick books on Greek philosophy. But very soon I became disillusioned with philosophy. I found that the business of the 'good' philosophers seemed mainly to refute the 'bad' philosophers. There didn't seem to be much else. The next-best thing seemed to be physics. Although physics does not address itself to the 'biggest' questions, still it does try to find out what the world is like. And it progresses. One generation builds on the work of another instead of simply overturning it. In my secondary school I was already beginning to get some idea that nature respects laws, like Newton's laws of motion. I remember a big disappointment when we started our course in Newtonian mechanics. It was in a room with models of steam engines all around. The teacher said, 'Next time we are going to play with the machines.' I thought he meant the steam engines. But it turned out that he meant things like levers. It was a great disappointment.

Because of the cost, the only university Bell could have considered going to after graduating from high school was the Queens

University in Belfast. But Bell had graduated from high school at sixteen, and the university would not admit anyone before the age of seventeen. So Bell looked for work. "I applied to be office boy in a small factory," he recalled, "some starting job at the BBC—things like that. But I didn't get any of the jobs I applied for. One told me that I was overqualified; another didn't tell me anything. It may be that I was resisting presenting myself as if I really wanted a job. I really wanted to continue on to the university, and the job I finally did get was in the university." It turned out that a laboratory assistant was needed in the physics department, and Bell got that job. "It was a tremendous thing for me," he said, "because there I met, already, my future professors. They were very kind to me. They gave me books to read, and in fact, I did the first year of my college physics when I was cleaning out the lab and setting out the wires for the students."

Among the professors that were especially helpful to him that year, Bell remembers Karl Emeleus and Peter Paul Ewald. Emeleus gave him two books to read: the classic freshman physics text *Mechanics, Molecular Physics, Heat and Sound* by Millikan, Roller, and Watson and an odd Victorian text on electricity and magnetism entitled *Elements of Electricity* by the British physicist J. J. Thomson. It was Thomson who in 1897 had identified the first subatomic particle—the electron. He measured its charge and mass, and for this work he was awarded the Nobel Prize in 1906. His work seemed to show that the electron was a particle—a motelike billiard ball. Ironically, Thomson was still alive when his son, G. P. Thomson, shared the 1937 Nobel Prize with C. Davisson for their independent experimental discoveries in the late 1920s that the electron can also act like a wave, one of the mysteries of the quantum theory. In any event, Bell found the Thomson *père* book exceedingly difficult. While its donor, Professor Emeleus, was a rather formal character, Ewald, a very distinguished crystallographer who, in Bell's words, "had been washed up on the shores of Ireland after the Nazis forced him to emigrate from Germany," was just the opposite. "He'd discuss anything," Bell noted. "He even declared that one of his assistants was mad."

By the time Bell graduated from Queen's College, in 1949, he had decided to make a career in theoretical physics. Ewald suggested that he continue his studies in graduate school with Rudolf—now Sir Rudolf—Peierls in Birmingham. Peierls was, and is, a physicist with an extraordinary breadth of interests. There is almost no branch of modern theoretical physics in which he has not made some very significant contribution. He is also a great teacher. While he was in Birmingham, several generations of young theoretical physicists came there to study with him. Some, like Freeman Dyson, even lived in the Peierls' house. Bell would have liked nothing better than to have gone off to Peierls except, as he put it, "By that time I had a very bad conscience about having lived off my parents for so long [Bell had lived at home while he was in college], and I thought I should get a job. So I did get a job, at the Atomic Energy Research Establishment at Harwell. There I went, and I found myself soon sent off to a substation at Malvern in Worcestershire." Bell was now twenty-one. While he returns to Ireland regularly to see his family (and in June 1988, in a single week, he was awarded honorary degrees from The Queen's College in Belfast, his alma mater, and from Trinity College in Dublin), he has not lived in Ireland since.

In the meantime, by an equally serendipitous route, Mary Bell had also arrived in Malvern. Mary was born in Glasgow into what Bell calls a "slightly more bookish family than my own." Her father began as a clerk in a shipyard and finally became a commercial manager specializing in the commerce of wood. Her mother was an elementary school teacher. Unlike Bell, Mary recalls being especially fond of arithmetic problems as a child. Fortunately, as it turned out, the school she went to—a comprehensive school with grades from elementary school through high school—was coeducational, which meant it offered a physics course. "With girls alone," Mary told me, "there very likely would not have been physics—other sciences, but not physics." She did very well in school and, as Bell is fond of reminding her, won many prizes for "general excellence." In 1941 she won a Bursary competition, which enabled her to go to the University of Glasgow, where she majored in mathematics and physics. "It was during the war," she reminds me, "and we all had to take courses in radio, circuits, and the like. At the end of my third year I was drafted to work in the radar lab at Malvern where, I must say, I didn't do a lot." After the war ended Mary finished her degree in Glasgow and was, like Bell, hired by the Atomic Energy Research Establishment and soon sent to Malvern, where she joined the same accelerator design group that had hired Bell.

As it happened, the Bells' arrival in the accelerator group at Malvern more or less coincided with a revolutionary breakthrough in the design of accelerators, known as the principle of "strong focusing," which was invented in the United States by Ernest Courant, M. Stanley Livingston, and Hartland Snyder, and independently by a Greek physicist-inventor named Nicholas Christofilos. A beam of charged particles that is being guided around the interior of an accelerator by electric and magnetic fields tends to try to "get away." Unless it is focused by these fields, the beam squirts off uncontrollably in all directions. For a while it appeared that this problem would be insurmountable for the new generation of machines being designed for the 1950s. But, as has happened so often in the accelerator business, there was an unexpected discovery that salvaged the situation. In this case Courant, Livingston, and Synder realized that if one used suitably varying electric and magnetic fields, rather than steady ones, as had previously been used to confine the beam, stability could be achieved. Bell told me that, prior to this discovery, a few designs had been generated in computer simulations that seemed to exhibit stability, but no one understood why.

Bell very quickly became an expert on the mathematics of strong focusing. Bell is best known among theoretical physicists for his abstruse work on the theory of elementary particles and quantum mechanics, so he is very pleased to have in his curriculum vitae such "engineering" items as the "Stability of Perturbed Orbits in the Synchrotron" written in 1954. ("Synchrotron" is the generic term for the kind of accelerator that uses strong focusing.) As young as he was, Bell became a consultant to the British delegation that was beginning the design studies that would lead, in 1959, to the construction of the first accelerator at CERN—the so-called Proton Synchrotron. It was a landmark in postwar European science, as it was the first major scientific project carried out after the war by the European community as a whole. Bell began consulting for the project in 1952. That, as it happened, was also the year that he got an unexpected job bonus—"a bolt from the blue," as he put it. It turned out that the Atomic Energy Establishment had the very enlightened policy of selecting some of their young people and sending them back to the universities for a year's study. "It was proposed to *me*," Bell recalled, still with a tone of surprise, "it didn't come from me, that I might go back to university for a year. And so I said 'fine,' and off I went to Birmingham and there I became a quantum field theorist." In the meanwhile the Bells had both moved from Malvern to Harwell and had gotten married.

Bell went off to Birmingham, and Mary remained at Harwell. "Our marriage was a thing of weekends for a while," he noted. In the British system at the time, there was in each department a single "professor" who functioned essentially as the chairman of the department. Peierls was *the* professor of theoretical physics at Birmingham, and he assigned the young people who came to work with him both problems to work on and junior members of the faculty to work with. Bell, having already worked as a professional physicist for three years, was somewhat older than the rest of the students. After noting that he did not intend to treat him as a "beginner," Peierls suggested a general area for Bell to look into and assigned him an advisor—a young British theorist named Paul Matthews. It was not clear whether or not the work was meant to lead to a degree, or even if a second year was possible. "I just didn't open up that question," Bell told me. "I thought a year was fine. I thought I was very lucky to get it. I just accepted that." Within a few months Bell discovered a very deep theorem in quantum field theory which is known as the TCP theorem. In this acronym, T stands for time reversal; C, for charge conjugation; and P, for parity. These are quantum-mechanical symmetries. The theorem states that the combined operation of these symmetries is a valid one even in theories where the individual symmetries may break down. It implies, among other things, that particles and antiparticles, such as electrons and positrons, have the same mass. "Unfortunately for me," Bell explained, "when I was writing that up there appeared not just a preprint but a *reprint* from Gerhard Lüders who had made the same discovery. So I was a year behind him. [Lüders's work was later generalized by Wolfgang Pauli and is often referred to as the 'Lüders-Pauli' theorem.] And I cannot exclude that there was some garbled rumor of Lüders's work that had reached Birmingham and that Peierls had asked me to look into that. Anyway, I thought I could make a paper out of my stuff and also part of a thesis."

When his year was up, Bell returned to Harwell to a new group that had been formed to do fundamental research in areas such as elementary-particle physics. Bell was able to do a second problem, which completed his thesis. "From time to time," he explained to me, "I had a remark to make to my old accelerator group. Mary was still there. I hadn't given much thought to my future. When I went to Harwell at the age of twenty-one I already had a tenure position. I didn't have to worry about anything. I just went along. So long as I was happy, I didn't think of going anywhere. But towards the end of the fifties I began feeling uncomfortable because there was a growing soul-searching at Harwell. What was that establishment supposed to be doing? They were not supposed to be doing nuclear weapons, though I believe now that there was some weapons work going on. Harwell had been set up to develop peaceful uses of atomic energy, but by that time, the nuclear power stations had already been built and other, more applied research establishments had grown up to do that kind of work. Harwell had sort of lost its sense of direction. Although I was in a very particular corner of Harwell, doing relatively fundamental work, this malaise was felt everywhere. It also began to look as if the fundamental research would be one of the things that would disappear in a reorientation of the establishment. So I started to think about going somewhere else." The "somewhere else" was CERN.

CERN—which stands for Conseil Européen pour la Recherche Nucleaire—had its formal beginnings in 1954. It is a consortium of thirteen member states: Austria, Belgium, Great Britain, Denmark, France, Greece, Italy, the Netherlands, Norway, Spain, Sweden, Switzerland, and West Germany. Neither the United States nor the Soviet Union is a member, although both Americans and Soviets work in the laboratory. No classified work of any kind is done at CERN. (A symptom of *glasnost* is the fact that the very best Soviet scientists are now coming to CERN for long visits *with* their entire families, something that was unheard of a few years ago.) In 1953 a referendum was held by the Canton of Geneva, and the voters ratified the decision of the Swiss government to give the nascent laboratory its site near the border. From the beginning, it was decided that CERN would have only a very small *permanent* staff of physicists compared with the visitors. For example, in the theory division, which now consists of about a hundred and forty people, less than ten percent are staff members. Typically, a staff member is offered a three-year contract with the understanding that, as a rule, a second three-year contract will be offered. Sometime during those six years a decision is made as to whether the staff member will be offered one of the few jobs with unlimited tenure.

There did not seem to be any university in Britain that could accommodate both Mary's interests in accelerator design and Bell's growing commitment to research in elementary-particle physics. CERN seemed ideal, except that they would be giving up the security of Bell's tenured job for three-year contracts. "I am amazed," Bell commented, "at how easily we left a tenured situation at Harwell and went to an untenured one. I remember that Mary's parents were a bit worried, but neither of us was worried. When we first got here in 1960 they already had the tradition that new people were more or less ignored and that you had to find your way to other people. Of course when I first showed up in the theory division they shook my hand and said 'Welcome,' but after that they left me alone. Nobody was coming into my office, or anything. I got quite lonely for the first months. It's my understanding that newcomers here can still feel like this for the first months. From time to time, we think that we must do something about it, and we have tried various schemes. But the trouble is that the staff members who are here are already so saturated that it is not easy to cold-bloodedly say I must go and spend time with this person because he or she is new. In any case, Mary and I settled into Geneva very easily. We got to know people we work with in CERN, and we didn't try to get to know many other people. So the fact that we were surrounded by Swiss rather than English and that they spoke French wasn't very important to us. And, of course, we were very happy with the new landscape around here. While we liked being in Berkshire, where Harwell is—there are a lot of beautiful things in Berkshire—the novelty of the high mountains here pleased us very much." Except for a very occasional sabbatical leave, the Bells have not left Geneva since 1960.

I once asked Bell whether during the years he was studying the quantum theory it ever occurred to him that the theory might simply be wrong. He thought a moment and answered, "I hesitated to think it might be wrong, but I *knew* that it was rotten." Bell pronounced the word "rotten" with a good deal of relish and then added, "That is to say, one has to find some decent way of expressing whatever truth there is in it." The attitude that even if there is not something actually wrong with the theory, there is something deeply unsettling—"rotten"—about it, was common to most of the creators of the quantum theory. Niels Bohr was reported to have remarked, "Well, I think that if a man says it is completely clear to him these days, then he has not really understood the subject." He later added, "If you do not get *schwindlig* [dizzy] sometimes when you think about these things then you have not really understood it." My teacher Philipp Frank used to tell about the time he visited Einstein in

Prague in 1911. Einstein had an office at the university that overlooked a park. People were milling around in the park, some engaged in vehement gesture-filled discussions. When Professor Frank asked Einstein what was going on, Einstein replied that it was the grounds of a lunatic asylum, adding, "Those are the madmen who do not occupy themselves with the quantum theory."

Max Planck, whom one may consider either the father or the grandfather of the quantum theory, depending on one's analysis of the history, was hardly a "lunatic." He was a conservative German, in the best sense of the term, from an ancient family of scholars, public servants, and lawyers. His father was a professor of law at Kiel, where Planck was born in 1858. Planck died in 1947 having lived with dignity in Germany during the Nazi regime. His son Erwin was executed by the Nazis after he took part in the July 1944 plot against Hitler.

What appealed to Planck about physics was the possibility of finding absolute laws that would retain their meaning, as he once wrote, "for all times and all cultures." One such law appeared to be the law governing what is known as "blackbody" or "cavity" radiation. A so-called blackbody can be made with a hollow, thin-walled cylinder of some metal such as tungsten. The walls of the cylinder are heated by, for example, passing an electric current through them. Radiation is then produced from the heated walls, and it collects within the hollowed cylinder. If a small hole is drilled in the cylinder, enough of this radiation gets out that one can measure its characteristics. (Incidentally, if the cylinder is kept at room temperature the hole looks perfectly black from the outside, since any radiation falling into it is trapped inside the cylinder—hence the name "blackbody.") In particular, blackbody radiation has a very characteristic distribution of wavelengths—colors. It turns out that this distribution depends only on the temperature to which the cylinder is heated, and not on the material of which the cylinder is made. Cavities made of, say, tantalum or molybdenum will exhibit the same blackbody spectrum as cavities made of tungsten, provided only that all the cavities are heated to the same temperature. This was demonstrated theoretically in 1860 by Planck's teacher at the University of Berlin, Gustav Kirchoff, using the newly discovered science of thermodynamics. He showed that if different materials had different blackbody spectra, one could construct a kind of perpetual motion machine by connecting the cavities together.

Kirchoff's result gave the blackbody spectrum the sort of absolute character that appealed to Planck. Moreover, in 1896, another German physicist, Wilhelm Wien, produced a simple formula that seemed to agree with both the thermodynamics and the empirical data. Planck set out to derive Wien's formula from something like first principles. Indeed, by 1899 he thought he had found such a derivation. However, by October 1900 Planck realized that his derivation was wrong. Furthermore, the Wien formula was beginning to break down experimentally when confronted with new data that embraced a wider range of wavelengths. Planck then made a sort of inspired guess as to what the correct formula was, and no sooner had he announced it than it was confirmed. He later wrote,

> The very next morning I received a visit from my colleague Rubens. He came to tell me that after the conclusion of the meeting [at which Planck had presented his formula], he had that very night checked my formula against the results of his measurements and found satisfactory concordance at every point . . . Later measurements, too, confirmed my radiation formula again and again—the finer the methods of measurement used, the more accurate the formula was found to be.

And so it has remained.

Planck tried to derive his formula. As he put it, "On the very day when I first formulated this law, I began to devote myself to the task of investing it with a real physical meaning . . . "

To understand what is at stake, we can imagine the walls of the blackbody cavity as being made up of atoms that are set into oscillatory vibrations when heated. These oscillators emit and absorb radiation, and when they emit as much radiation as they absorb, an equilibrium situation is produced that is precisely that of a blackbody. In classical physics, there are no restrictions on how much energy an oscillator can emit or absorb in a single transaction; there are no minimal units. What Planck discovered was that he could derive his formula if he allowed the emission and absorption of radiation to take place only in discrete units—lumps of energy. (The use of the term "quanta" for these energy units was introduced by Einstein in 1905.) In Einstein's homey image, it was like selling beer from a keg in pint bottles. As historians of science such as Thomas Kuhn have noted, the derivations Planck gave in his papers of 1900 and 1901 are sufficiently obscure that one cannot be sure whether Planck understood that he had made a radical new assumption or whether he thought he was still doing classical physics.

Fortunately, Planck made an unjustified step in his derivation. If he had done it conventionally it would not have led to the Planck formula (nor to the Wien formula) but to the expression derived by Lord Rayleigh in 1905 that classical physics inexorably predicts. This expression, called the Rayleigh-Jeans law because it was slightly amended by the British astrophysicist James Jeans, agrees with the Planck law for long wavelengths, but then leads to disaster. It predicts that the total energy in the cavity is *infinite*, a nonsensical proposition that became known as the "ultraviolet catastrophe." In short, by 1905 it was clear, if only to Einstein, that classical physics had broken down.

A hint of Einstein's state of mind at this realization is contained in his letter to Habicht that I quoted earlier. Of his 1905 papers, including the one on the theory of relativity, which modified our notions of space and time, it is only the paper on the blackbody spectrum that Einstein described as "revolutionary." We will shortly see why.

As in so many other aspects of physics, the quantitative attempts to study the nature of light begin with Newton. Newton spent at least as much time thinking about light as he did about gravitation; the same can be said about Einstein, although both men are best known for their theories of gravitation. Newton believed that light was a particlelike projectile emitted by a radiating object. In the English edition of his book *Optiks*, published in 1717, he described rays of light as being composed of "very small Bodies emitted from shining Substances." However to explain some of his observations he supplemented this simple particle picture with the notion that these particles could incite wavelike disturbances in an all-pervasive medium that became known as the ether. These disturbances he called "Fits of easy Reflexion and easy Transmission." On the other hand, his less well-known but almost equally great contemporary, the Dutch physicist Christian Huygens, maintained that light was nothing but a wave disturbance in the ether. He worked out the mathematics of such wave propagation in a very sophisticated way that is still used.

A pure wave theory of light makes predictions about optical phenomena that appear to be incompatible with any particle description whatsoever. For example, according to the wave theory objects do not cast sharp shadows. Waves curl around objects, so their shadows are not sharply defined—a phenomenon known as diffraction. A familiar example is a breakwater in a harbor. Water waves spread out behind a breakwater; a stream of particles, on the other hand, confronting such an obstacle would simply be deflected by it, and the obstacle would act like an ineluctable barrier to their transmission. At the heart of the distinction is the phenomenon of interference. When two waves encounter each other they do not simply bounce off one another like colliding billiard balls. Rather, the two wave forms combine to produce a new wave pattern. Separate wave disturbances can reinforce each other to produce amplifications, or they can interfere destructively to produce areas of little or no wave activity. In the case of light waves these would be dark regions; in the case of sound waves these would be places where the sound is muted—a problem that besets the designers of concert halls.

Both the particle and wave theories of light had their advocates until the beginning of the nineteenth century. However, in 1801 the British natural philosopher Thomas Young did an experiment that seemed to settle the matter once and for all in favor of the wave theory. He reflected sunlight from a set of parallel grooves cut into glass. He observed that the light coming through these grooves produced a pattern of alternating light and dark fringes, indicating interference effects. There was no doubt in Young's mind what this meant; he concluded bluntly that "Radiant Light consists in Undulations of the luminiferous ether." All of nineteenth-century optics and electromagnetic theory was built on this discovery, including the grand synthesis created in the second half of the century by the Scottish physicist James Clerk Maxwell, whose theory of electricity and magnetism was to that subject what Newton's theory of gravity was to its. It was against this apparently overwhelming preponderance of evidence in favor of the wave theory of light that Einstein's paper of 1905 was written.

In contemplating the papers Einstein wrote in 1905, I often find myself wondering which of them is the most beautiful. It is a little like asking which of Beethoven's symphonies is the most beautiful. My favorite, after years of studying them, is Einstein's paper on the blackbody radiation. The paper has the mind-numbing German title "*Über einen die Erzeugung und Verwandlung des Lichtes betreffenden heuristischen Gesichtspunkt*"—in English, "Concerning a Heuristic Point of View about the Creation and Transformation of Light." It begins with a brief discussion of the difference between gases and what Einstein calls "other ponderable bodies," and light. The former appear to be made out of particulate elements—Einstein mentions atoms and electrons—while light is represented as a continuous medium. Mincing no words, Einstein states that the radiation in a blackbody cavity, and elsewhere, *pace* the wave theory, may in fact have a particulate character. He writes, "According to the presently proposed assumption the energy in a beam of light emanating from a point source is not distributed continuously over larger and larger volumes of space but consists of a finite number of energy quanta, localized at points of space which move without subdividing and which are absorbed and emitted only as units."

In terms of Einstein's previously quoted homey image, not only was the beer bought and sold in pint containers, but the beer in the keg could exist only in pint-container units.

This having been said, in the second section of the paper Einstein produces the formula for the blackbody distribution that Planck should have gotten if he had done the calculation correctly in classical physics—the Rayleigh-Jeans law. Einstein arrived at it independently. He then points out that this law leads to an absurdity, namely an infinite total energy in the cavity—the ultraviolet catastrophe.

It is in the fourth section of the paper that the true novelty begins. Having come to the conclusion in the first part of the paper that classical physics can correctly describe only the long-wavelength parts of the spectrum—the "graver modes," in Lord Rayleigh's wonderful terminology—Einstein focused his attention on the opposite end of the spectrum, the short wavelengths, which must contain the new physics. Here the spectrum is described by the Wien law, so, Einstein reasoned, the new physics must be hidden there.

Part of being a great scientist is to know—have an instinct for—the questions *not* to ask. Einstein did not try to *derive* the Wien law. He simply accepted it as an empirical fact and asked what it meant. By a virtuoso bit of reasoning involving statistical mechanics (of which he was a master, having independently invented the subject over a three-year period beginning in 1902), he was able to show that the statistical mechanics of the radiation in the cavity was mathematically the same as that of a dilute gas of particles. As far as Einstein was concerned, this meant that this radiation *was* a dilute gas of particles—light quanta. But, and this was also characteristic, he took the argument a step further. He realized that if the energetic light quanta were to bombard, say, a metal surface, they would give up their energies in lump sums and thereby liberate electrons from the surface in a predictable way, something that is called the photoelectric effect. Some evidence that this was true already existed in 1905, but it was not very convincing. It became convincing in the next decade, and when Einstein won the Nobel Prize in 1922 it was for this prediction and not for the theory of relativity.

However, it must not be imagined that there was an epiphany among physicists after Einstein published this paper in 1905. In the first place, not many physicists were even interested in the subject of blackbody radiation for at least another decade. Kuhn has done a study that shows that until 1914 less than twenty authors a year published papers on the subject; in most years there were less than ten. Planck, who was interested, decided that Einstein's paper was simply wrong. He wrote a celebrated letter of recommendation for Einstein for his admission to the Royal Prussian Academy of Sciences in 1913, in which he said of Einstein, "that he may sometimes have missed the target in his speculations, as for example, in his theory of light quanta, cannot really be held against him." One even wonders about Einstein's attitude toward this work. Three months after he finished the paper on the light quanta, he submitted his paper on relativity to the *Annalen der Physik*—the same journal that published the quantum paper. In the relativity paper Einstein also deals with the theory of light; indeed, the propagation of light signals plays a central role. For the purposes of this paper light is treated as a wave phenomenon. There is not a hint that it might have, under certain circumstances, a particle aspect. There is something almost schizophrenic about this separation of ideas in the two papers. It is a schizophrenia that haunted Einstein, and has haunted the quantum theory ever since.

Two developments brought the quantum theory to the center of physics. The first began in 1907, when Einstein wrote a paper that founded the quantum theory of solids, what we now call solid-state or condensed-matter theory. It is this discipline that lies behind much of modern technology, from the superconductor to the transistor. Einstein was concerned with how solid bodies absorb heat. He imagined that the atoms became agitated and oscillated when the substance absorbed heat. He applied the same quantum rules to these oscillators as Planck had used in his derivation of the blackbody law. This led to a new theory of heat absorption, which provided an explanation of some heretofore puzzling experimental results. That, in turn, aroused the interest of a community of scientists who had previously ignored the quantum theory. For example, in 1910 the great German physical chemist Walther Nernst made a special trip to Zurich, where Einstein was teaching, to discuss these matters with him. Nernst's public endorsement of Einstein's work influenced many other scientists to begin to take it seriously. But the real breakthrough came in 1913 with Bohr's invention of the "planetary" model of the atom. The Bohr atom, with its electrons in circular orbits around a central nucleus, has become one of the defining pictorial images of the atomic age.

Not until 1909 was it known that the atom had a nucleus. That year the New Zealand–born experimental physicist Ernest Rutherford directed his younger colleagues at Manchester University to do a series of experiments in which they allowed helium nuclei, which are emitted with substantial energies from certain naturally occurring radioactive substances, to impinge on a sheet of gold foil some fifty thousandths of an inch thick. Rutherford expected that these projectiles would pass directly through the foil. Instead, some of them bounced backward as if they had struck something hard in the interior of the gold atoms. Rutherford later described his astonishment. He wrote, "It was quite the most incredible event that has ever happened to me in my life. It was almost as incredible as if you fired a 15 inch shell at a piece of tissue paper and it came back and hit you."

What the helium projectile had hit was the gold-atom nucleus—the tiny, and incredibly dense, part of the gold atom where most of its mass is located. (We now know that the nucleus is made up of protons and neutrons and that it is surrounded by a cloud of relatively light, electrically charged electrons.) At the time of this discovery, Bohr was still a student at the University of Copenhagen. After taking his doctoral degree in 1911, he went to Cambridge, hoping to work on electron theory with the discoverer of the electron, J. J. Thomson. It turned out that Bohr and Thomson never quite hit it off, and after a short time Bohr decided to go to Manchester with Rutherford. They made an odd but very well-suited pair—the shy Bohr and the enormously enthusiastic and totally self-confident Rutherford. At Manchester, Bohr began the work that created the modern atom.

The basic problem that Bohr confronted was, what kept the atom stable, and why, when it was energized by electrical discharges or otherwise, did it return to stability by emitting beautiful patterns of spectral light. Classical physics taught that the electrons moving around the nucleus should radiate. As the electrons radiated, they lose energy and should then, according to classical theory, fall into the nucleus. Moreover, the classical energy radiated this way bears no resemblance to the beautifully ordered spectral lines given off in reality by excited atoms. Getting those harmonic patterns from a gas of chaotically collapsing electrons seemed about as likely as dropping a piano from a fourth-floor window and having it play, as it hit the sidewalk, Beethoven's "Moonlight Sonata."

To explain both the atomic stability and the spectral regularities, Bohr made the radical assumption that the electrons outside the nucleus were allowed to occupy only certain select orbits (these came to be known as the "Bohr orbits") and no others. The Bohr orbit with the lowest energy, the so-called ground state, was, he said, absolutely stable. On the other hand, electrons in the orbits with higher energies, the "excited" states, could make spontaneous transitions to the ground state with the emission of light quanta whose energies were determined by the energy differences of the electrons in the various Bohr orbits. No explanation was offered for these rules, and no accounting was given for what the electron did while it was making such a "quantum jump." What persuaded everyone, including Einstein, who called the Bohr atom "one of the greatest discoveries," that Bohr had done something of fundamental importance was that by using his rules, Bohr was able to derive a mathematical formula that gave the frequencies of the spectral lines in hydrogen to great accuracy. For the next decade theoretical physicists tried to bring Bohr's rules into some general context and to apply them to increasingly more complicated configurations of electrons. This enterprise now goes under the rubric the "old" quantum theory. It was swept away by the "new" quantum theory, which was developed during a five-year period beginning in 1923. That is the version of the quantum theory that has endured to the present day.

The first step in its creation came from a totally unexpected quarter—the Prince Louis de Broglie of Paris, who was born in 1892. It was a case of sibling rivalry. The older brother, Maurice, had become so enamored of physics while serving in the French Navy that he contemplated resigning his commission in order to do research. This scandalized his family. His grandfather noted that science was "an old lady content with the attractions of old men." As a compromise, a laboratory was set up for him in a room in the family mansion in Paris. Maurice de Broglie became a first-rate x-ray spectroscopist, and a generation of French experimental physicists after the First World War was trained in de Broglie's mansion. The activity attracted his younger brother, Louis, who decided to have an equally distinguished career in physics.

In the early 1920s the brothers worked together studying various x-ray phenomena. But in 1923 Louis de Broglie had an idea (it would become his Ph.D. thesis) that transformed modern physics. He was familiar with Einstein's early papers in which light had been given a particulate nature, and largely for reasons of symmetry, he proposed that particles such as electrons, which had previously been thought of as sort of billiard balls in miniature, should be given a wave nature. He referred to these waves as "fictitious," since their relationship to the particulate electrons was unclear. But he was, nonetheless, able to give a kind of "explanation" of the location of the Bohr orbits by noting that their circumferences were just large enough that a whole number of electron wavelengths would fit into a given circumference. He pointed out that this speculation could be tested with the sort of diffraction experiments that Young had used to demonstrate the wave nature of light. However, de Broglie predicted that the electron waves would have wavelengths something like a thousand times smaller than those of visible light. Hence different techniques would have to be used to see the diffraction patterns.

These experiments were carried out in 1927 by C. Davisson and L. Germer in the United States using crystals and by G. P. Thomson in Britain using thin films. (The electron microscope, developed in the 1930s, produces its high magnification by exploiting the shortness of the de Broglie waves.) At the time de Broglie made his speculation he did not have his doctorate, so he submitted this work as his Ph.D. thesis. His thesis advisor, Paul Langevin, was somewhat uncertain as to what to make of it, so he sent a copy of the thesis to Einstein, who immediately grasped its potential importance. It is even possible that he had been thinking along similar lines. Einstein, in turn, sent a letter about this to the Dutch physicist Hendrik Lorentz, whom Einstein most admired among the physicists of the generation that preceded his own. Einstein wrote, "A younger brother of . . . de Broglie has undertaken a very interesting attempt to interpret the Bohr-Sommerfeld quantum rules [this was the business of fitting the electron waves around the Bohr orbits] . . . I believe it is the first feeble ray of light on this worst of our physics enigmas. I, too, have found something which speaks for his construction." De Broglie won the Nobel Prize in 1929 for the work he did for his doctoral thesis.

The next step was taken in 1925 by the young German physicist Werner Heisenberg. He too appears to have been inspired by Einstein, who Einstein hovers over this entire subject like some sort of magisterial ghost. Heisenberg had absorbed what he thought was the philosophical lesson of the theory of relativity, namely that physics gets itself into trouble when it becomes based on metaphysical abstractions instead of concepts with direct links to experimental procedures. In the relativity theory, Einstein had replaced the metaphysical notions of "absolute" space and time by experimental procedures involving clocks and metersticks. Indeed, from this point of view space and time have no other significance than relationships measured by clocks and metersticks.

Heisenberg decided that Bohr's emphasis on atomic electron orbits—despite their pictorial appeal—had been misplaced. No one ever observed an electron actually circulating in an atomic orbit. What one did observe was the radiation emitted by these electrons as they lost energy—the atomic spectra. The orbits were, from this point of view, a gratuitous remnant of classical physics. Heisenberg decided, therefore, to focus directly on the computation of the wavelengths of these spectral lines even if this meant giving up the visualization in terms of electrons circulating in orbits.

To this end, Heisenberg created a mathematical scheme known as matrix mechanics. Matrices, which were first studied in the nineteenth century, obey unconventional multiplication laws. The product of A times B is not, in general, equal to the product of B times A, whereas for ordinary numbers, $5 \times 3 = 3 \times 5 = 15$. Heisenberg, who knew nothing about matrices, reinvented the subject by studying the properties of the atomic spectra.

Once he had his matrix mechanics, Heisenberg was able to reproduce all the results of the old quantum theory and more. It was the first example of a kind of Faustian bargain quantum theorists were to make with the spirit of visualization—namely, one would be allowed to predict experimental results with very high accuracy provided that one did not ask for a visualization of the phenomena that went beyond the rules themselves.

Einstein followed Heisenberg's work with interest, but it was the next step, taken by Heisenberg's older contemporary, Erwin Schrödinger, that at least initially aroused his enthusiasm. Schrödinger, who was born in Vienna in 1887, was a fascinating man. I met him a few months before he died in 1961. All the inventors of the quantum theory, as it happened, were men of very broad culture, perhaps attributable in part to their European *gymnasium* educations, but even in this group Schrödinger stood out. He read very widely in a variety of languages, ancient and modern. He was a scientific polymath with a deep interest in Eastern religions. He was also a rather romantic figure who wrote poetry. I was told by Philipp Frank that when Schrödinger appeared in 1939 at the Institute for Advanced Study in Dublin, where he had been offered sanctuary during the war, he did so with what Professor Frank referred to as two "wives." (This was the least of it. Schrödinger had several mistresses, with whom he fathered at least two daughters.) Schrödinger's marvelous book *What is Life?*, written in Dublin in 1944, inspired a whole generation of young scientists such as Francis Crick to take up biology. In 1918 Schrödinger thought he might get a chair of theoretical physics at the provincial university of Czernowitz. He was then thirty-one. As he later wrote, "I was prepared to do a good job lecturing on theoretical physics . . . but for the rest to devote myself to philosophy . . . My guardian angel intervened: Czernowitz soon no longer belonged to Austria. So nothing came of it. I had to stick to theoretical physics, and, to my astonishment, something occasionally emerged from it." In 1926, "wave mechanics" emerged from it.

He later wrote that in discovering the equation that bears his name—which is the heart and soul of the quantum theory—he had been "stimulated by de Broglie's thesis and by short but infinitely far seeing remarks by Einstein." The Schrödinger "wave equation" describes how the de Broglie (or what Schrödinger insisted on calling the "Einstein–de Broglie") waves propagate in time and what their form in space is for all physical systems, from the electrons in an atom to the neutrons and protons in uranium. It is the basic tool of the modern theoretical physicist, to say nothing of the chemist and the electronics engineer.

None of this was foreseen in 1926. The first reactions to wave mechanics were mixed, to put it mildly. Einstein was extremely enthusiastic and wrote to Schrödinger, "The idea of your article shows real genius." Heisenberg, on the other hand, was severely critical—polemical, in fact—of wave mechanics. For him it was a return to the kind of visualization of quantum phenomena in terms of classical pictures that went beyond the empirical data.

It looked for a while as if there were two different quantum theories with radically different mathematical and philosophical underpinnings. That perception changed for two reasons. In the first place, Schrödinger and, independently, the British theoretical physicist P.A.M. Dirac were able to show that wave mechanics and matrix mechanics were simply two equivalent ways of representing a common underlying mathematical structure, now simply called quantum mechanics. In the second place, it soon became evident that Schrödinger's waves were no less abstract than Heisenberg's matrices. The idea that they could be interpreted like, say, water waves on a pond, the sort of thing that appealed to Einstein, was an illusion.

As I have mentioned, when de Broglie first introduced his waves it was not clear how they were to be related to an object like an electron. Was the electron the wave, or did the wave attach itself to the electron and guide it, or what? De Broglie and Einstein certainly had the idea that the waves were physical objects that existed in ordinary, tangible three-dimensional space—the kind we exist in. Einstein spoke of "*Führungsfelder*" ("pilot waves") that guided the electrons and other particles. But it soon became clear that any such simple picture was impossible.

To see the sort of thing that goes wrong, take as an example a free electron, one that is not confined to an atom. We know that we can measure its position as accurately as we like. This means that the electron can be confined by this measurement to as small a region of space as we want. To describe this in a wave picture, it would mean that the wave describing the electron would have to be squashed down into this small region. This is quite possible, but then we can ask what happens to this squashed wave after the measurement. This question is answered by studying the Schrödinger equation. The results are dramatic. Suppose, for the sake of argument, we confine the electron to a region something like the size of an average atom. Then, once the measurement is over, the Schrödinger equation implies that the electron wave will spread out again very rapidly. Indeed, under the conditions I have just described, the electron wave would spread out over the entire solar system in about four days! If the electron were really a three-dimensional wave, this would mean that it had now become as big as the solar system. Clearly something was wrong.

A way out was suggested in a four-page paper published by the German physicist Max Born. Born, recognizing these paradoxes, argued that the solutions to the Schrödinger equation—which are called wave functions—should not be thought of as real waves attached to particles, but rather as mathematical artifacts to be used, (to take the example we have discussed) to compute the *probability* of finding the electron in a certain region of space. (In the body of his paper Born suggests that the wave function itself is the probability, but in a footnote he remarks that it really should be the *square* of the wave function that represents the probability. That, in fact, is the correct connection.) From this point of view, the bizarre behavior of the rapidly spreading electron described above is interpreted to mean that, if we measure the electron's position accurately by confining it to a minute region, then at later times there is some probability of finding the same electron at a far distant place. According to this interpretation the wave function does not tell us where the electron is, but only where it is likely to be.

Born was fully aware what a revolutionary step this was. In his paper he writes, "Here the whole problem of determinism comes up." He briefly raises the possibility that in the future one might find some "inner properties of the atom," as he calls them (later the term "hidden variables" would be introduced for the same idea), whose discovery would restore the deterministic character of the theory. But, he concludes, "I myself am inclined to give up determinism in the world of atoms."

A year later Heisenberg quantitatively spelled out the limits to determinism in quantum mechanics in what became known as the Heisenberg "uncertainty principle." A useful example is again provided by the attempt to measure the position of an electron. It is all very well to state, in the abstract, that one has performed an accurate measurement of the electron's position, but how would one go about doing so in practice? In his paper, Heisenberg proposes using a microscope. Light shining on the electron will be reflected from it and will pass through the lenses of the microscope. To increase the accuracy, one must decrease the wavelength of the light. But the shorter the wavelength, the more energy and momenta the light quanta carry. When these energetic light quanta hit the electron, they knock it for a loop. The future position of the electron, along with its momentum, becomes more indeterminate the more accurately we insist on measuring the electron's present position. Position and momentum are known in quantum mechanics as "conjugate quantities"—the more accurately one quantity is known, the less accurately the other is known.

A more precise statement of this conjugate uncertainty is given in terms of the Heisenberg relation

$$\Delta p \, \Delta q \geq \hbar.$$

Here Δp and Δq stand for the uncertainties in momentum and position, respectively. The quantity $\hbar$ is related to Planck's constant h by the equation

$$\hbar = h/2\pi = 1.054 \times 10^{-27} \text{ erg} \cdot \text{seconds.}$$

While this number involves dimensional quantities such as ergs (energy) and seconds (time), it is extremely small in the sense that this uncertainty relation plays no role in macroscopic physics. In a game of billiards, for example, we can ignore the Heisenberg relation. However, it is h that measures the departure from classical physics. If it had been zero, there would have been no quantum world.

If quantum mechanics is right, there is no way to get around the uncertainty principle. The reason that the electron's probability wave spread so much after we confined it, Heisenberg would argue, is that its momentum became almost completely indeterminate. In a manner of speaking, it headed off in all directions.

There is something amusing about Heisenberg's use of the microscope in his paper. A few years before it was written, he had taken his Ph.D. qualifying oral examination and had been asked how a microscope works. He could not answer the question and nearly failed the oral. In fact, in his 1927 paper he got it wrong again. In an addendum he thanks Bohr for straightening the matter out.

In 1926, even before Heisenberg's paper, Einstein abandoned the quantum theory. We know this because on December 4, 1926, he sent Born a letter that contains his most often quoted appraisal of the theory. He wrote, "Quantum mechanics is certainly imposing. But an inner voice tells me that it is not yet the real thing. The theory says a lot, but does not really bring us any closer to the secret of the 'old one' [Einstein's affectionate term for God]. I, at any rate, am convinced that *He* is not playing at dice."

Born was profoundly disturbed by this letter. When the letters were collected and published in 1971, Born wrote a commentary. Of this letter he commented, "Einstein's verdict on quantum mechanics came as a hard blow to me; he rejected it not for any definite reason, but rather referring to an 'inner voice' . . . It [the rejection] was based on a basic difference of philosophical attitude, which separated Einstein from the younger generation to which I felt that I belonged, although I was only a few years younger than Einstein."

Born also comments on the matter of the Nobel Prizes. Heisenberg, who was Born's junior colleague in Göttingen, received his in 1932, while, for reasons known only to the Swedish Academy, Born had to wait until 1954 to receive his for the work done in 1926, which had been the essential step leading to Heisenberg's work. Born writes, "The fact that I did not receive the Nobel Prize in 1932 together with Heisenberg hurt me very much at the time, in spite of a kind letter from Heisenberg. I got over it, because I was conscious of Heisenberg's superiority." There are things one never gets over, and as the rest of the commentary shows, this was one of them. Born goes on, "It is not surprising that this acknowledgment [Born's Nobel Prize] was delayed for twenty-eight years, for all the great names of the initial period of the quantum theory were opposed to the statistical interpretation: Planck, de Broglie, Schrödinger, and, not least, Einstein himself. It cannot have been easy for the Swedish Academy to act in opposition to voices which carried as much weight as theirs; there I had to wait until my ideas had become the common property of all physicists. This was due in no small part to the cooperation of Niels Bohr and his Copenhagen school, which today lends its name almost everywhere to the line of thinking I originated."

By the early 1930s, in fact, the statistical interpretation of quantum theory—the wave function as probability—was the "common property of all physicists." Very little attention was being paid to the objections of the grand old men of the previous generation. The Swedish Academy has certainly done a number of strange things when it comes to the awarding of Nobel Prizes, and not giving one to Born until twenty-eight years after his work was done is certainly one of them. Typically, Einstein sent Born a prompt letter of congratulations for the awarding of his "strangely belated" Nobel Prize. He adds, "It was your . . . statistical interpretation of the description which has decisively clarified our thinking. It seems to me that there is no doubt about this at all, in spite of our inconclusive correspondence on the subject." With typical Einstein humor he notes, "And then the money in good currency is not to be despised either, when one has just retired."

When, in his Reith lectures, Oppenheimer referred to a re-creation of history that "would call for an art as high as the story of Oedipus or the story of Cromwell," there is little doubt that he had in mind the dialogues that took place about the quantum theory between Einstein and Bohr. It is said that, once these began in the 1920s, Einstein was never very far away from Bohr's thoughts. Some sense of this is conveyed by an experience that the physicist Abraham Pais had in 1948 when he was helping Bohr prepare his account of these dialogues that was published in 1949 in the collection *Albert Einstein: Philosopher Scientist*. Bohr's method of composition—writing was always an agony for him—was to find a younger colleague and dictate while pacing up and down. In 1948 Bohr was a visitor at the Institute for Advanced Study in Princeton. Einstein had a large office there and an adjacent office for an assistant. Since Einstein did not like the large office, he had moved into the assistant's office, so the large office was available for Bohr. Bohr invited Pais into his office. Pais recounts, "He then asked me if I could put down a few sentences as they would emerge during his pacing. It should be explained that, at such sessions, Bohr never had a full sentence ready. He would often dwell on one word, coax it, implore it, to find the continuation. This could go on for many minutes. At that moment the word was 'Einstein.' There Bohr was, almost running around the table and repeating 'Einstein . . . Einstein . . . ' It would have been a curious sight for someone not familiar with Bohr. After a little while he walked to the window and gazed out, repeating every now and then, 'Einstein . . . Einstein . . . '

"At that moment the door opened very softly, and Einstein tiptoed in.

"He beckoned me with a finger on his lips to be very quiet, his urchin smile on his face. He was to explain a few minutes later the reason for his behaviour. Einstein was not allowed by his doctor to buy any tobacco. However, the doctor had not forbidden him to steal tobacco, and this was precisely what he set out to do now. Always on tiptoe, he made a beeline for Bohr's tobacco pot, which stood on the table at which I was sitting. Bohr, unaware, was standing at the window, muttering, 'Einstein . . . Einstein . . . ' I was at a loss what to do, especially because I had at that moment not the faintest idea of what Einstein was up to.

"Then Bohr, with a firm 'Einstein,' turned around. There they were, face to face, as if Bohr had summoned him forth. It is an understatement to say that for a moment Bohr was speechless. I myself, who had seen it coming, had distinctly felt uncanny for a moment, so I could well understand Bohr's own reaction."

Then, Pais reports, "A moment later the spell was broken when Einstein explained his mission. Soon we were all bursting with laughter."

Although Bohr first met Einstein in Berlin in 1920, it was not until 1927 that the discussions began in earnest. Ever since 1911, the Belgian industrialist Ernest Solvay had been funding a series of international physics conferences. Einstein had been a frequent participant, and he and Bohr were both in attendance at the 1927 conference, which was held in Brussels. From this discussion, which I will describe shortly, one can see that Einstein's objections to the quantum theory, as they began to emerge in 1927, go far beyond the fact that the quantum theory resorts to probabilities. However, the phrase "*He* is not playing at dice," which Einstein used to describe God in his letter to Born, is the one that has stuck. As the postwar Born-Einstein letters make clear, even Born did not understand that "dice-playing" was not what ultimately bothered Einstein. Rather, it was the idea that there was nothing beyond "dice-playing," that quantum theory was the *complete* description of reality.

The discussions of 1927 focused on what has come to be known as the "double-slit experiment." I have described how in 1801 the British natural philosopher Thomas Young "proved" that light consisted of waves: sunlight, falling on a grating made up of parallel grooves cut in glass, makes patterns on the other side—regions of light and dark. To see how this comes about in the wave theory, let us simplify the grating so that it consists of only two parallel slits through which the light can pass. If we cover one of the slits, so that all the light goes through the other, then what emerges is a diffuse beam of light with no especially interesting features. If we now open the second slit, the diffraction pattern appears. To understand this, let us concentrate on some specific location—any location—beyond the slits. In general, this location will be farther from one slit than the other. Hence the waves from the two slits may arrive out of phase. The crests from one wave will not necessarily arrive in synchrony with the crests from the other. This is what produces the interference. For some locations there will be synchrony, and at them the light waves will reinforce each other, making a bright spot in the interference pattern. It is absolutely essential to have waves arriving from both slits to see anything interesting.

This is how Young reasoned. But Einstein and Bohr knew that light in some circumstances appears to be made up of particles. How can this be reconciled with Young's result? To make the situation as dramatic as possible, suppose, with Einstein and Bohr, we imagine reducing the intensity of the beam of light to such a low level that only one light quantum at a time is available to go through the slits—what will happen? We may suppose that we have put some sort of photosensitive screen behind the slits to detect the quanta as they come through, one after the other. Each time a quantum hits a spot on the screen it blackens it, say. At first nothing interesting will happen. The quanta that get through will blacken a few apparently random spots on the screen. But, in the course of time, something remarkable happens. The spots on the screen will build up to make just the interference pattern that Young observed. From the point of view of a classical physicist this is quite mad! A classical physicist would argue that if the light quantum is really a particle it would have a well-defined trajectory that would take it through one slit or another—not both at once—which would seem to be what is required if one is going to produce interference fringes.

Einstein, with his years of experience in the patent office, even produced an imaginary modification of the double-slit device that enabled it to measure the photon's trajectory. Einstein imagined mounting the slits on springs so that they could jiggle when the photon passed through. That way one could make a definite statement that the photon had gone through one slit or another on its way to the screen. If this modified device still produced an interference pattern, the quantum theory would be inconsistent. The particle and the wave properties of light, which are incompatible with each other, would have been forced to manifest themselves together in a single experimental arrangement. But Bohr pointed out that Einstein's reasoning had neglected the Heisenberg uncertainty principle and therefore was not a valid criticism of the quantum theory. It has assumed that the jiggling the photon produces—which is a transfer of momentum from the photon to the structure holding the slits—can be measured without limitation from the uncertainty principle. But, as Bohr pointed out, if the momentum is measured accurately enough to determine the trajectory, then the uncertainty principle implies that the position of the jiggled slits is sufficiently uncertain so that no interference fringes can be produced.

This was an extremely important observation. It was an example of something that Bohr elevated to a general principle that he called the "principle of complementarity." In the case at hand, the complementary characteristics are the particle and wave aspects of light. The principle states that, in this case, any experiment designed to reveal the particle aspect of light cannot reveal its wave aspect and vice versa. Bohr enunciated this principle after analyzing a variety of examples like the double-slit experiment and seeing that in each example the limitations on measurement imposed by the quantum theory—the uncertainty principle—are just such as to prevent the theory from being self-contradictory.

Bohr was so impressed by this revelation that he began applying it to areas outside physics. He thought he saw echoes of this idea of complementary qualities in ethics (truth as opposed to justice), in psychology (thoughts as opposed to sentiments), in biology (mechanism as opposed to vitalism). He wrote and lectured extensively about these matters, but he had such difficulty in reducing his ideas to words that one wonders if any of it, the quantum mechanics aside, will survive.

The great value of this discussion for the quantum theorists was that it forced them to confront, head on, the quantum-mechanical nature of reality. Is light *really* a particle or is it *really* a wave? The quantum theorist would answer it *really* is neither. Light is just what it is revealed to be, in some given experimental arrangement. It was this aspect of the quantum theory, and not the dice-throwing, that profoundly bothered Einstein. There seemed to be a loss of objective reality. Pais recalls discussing these matters with Einstein on walks in Princeton. During one walk, Pais remembers, "Einstein suddenly stopped, turned to me and asked whether I really believed that the moon exists only when I look at it."

The next confrontation between Einstein and Bohr took place at the Solvay Congress of 1930, also held in Brussels. Einstein had prepared a little surprise for Bohr. This was an imaginary device that Einstein thought violated the uncertainty principle involving energy and time. The more precisely an energy is measured, the less certain is the time at which that measurement takes place. Einstein's imaginary contraption consisted of a box containing a clock along with some radioactive element. The clockwork is arranged so that it opens a hole in the box for a fraction of a second, just long enough to allow some radiation out, but short enough so that the time when this happens can be precisely determined. One can imagine that the box is attached to a scale that can measure its weight before and after the release of the radioactivity. In this way one could, Einstein claimed, determine the energy of the released radioactivity, which corresponds to the weight lost by the box, along with the precise time at which the energy was released—a violation of the energy-time uncertainty principle.

Bohr was stunned. His reaction was later described by Léon Rosenfeld, a Belgian physicist who was a close associate of Bohr. Rosenfeld writes, "It was quite a shock for Bohr . . . he did not see the solution at once. During the whole evening he was extremely unhappy, going from one to the other and trying to persuade them that it couldn't be true, that it would be the end of physics if Einstein were right; but he couldn't produce any refutation. I shall never forget the vision of the two antagonists, Einstein, a tall majestic figure, walking quietly, with a somewhat ironical smile, and Bohr trotting near him, very excited . . . The next morning came Bohr's triumph."

During the night, Bohr had realized that Einstein had left out of his argument an important consideration involving his own general theory of relativity and gravitation. A clock in a gravitational field runs more slowly than a clock not influenced by gravity. (This remarkable prediction of the theory has now been confirmed experimentally by very accurate atomic clocks flown in satellites and airplanes.) This effect plays a role in Einstein's imaginary device because, as Bohr realized, the position of the clock is somewhat uncertain owing to the uncertainty principle. Since the gravitational field varies from place to place, this affects the rate of the clock, which is therefore uncertain to the extent that the clock's location is uncertain. When that is taken into account, the uncertainty relation between energy and time is restored.

The fact that Einstein had not realized this consequence of his own theory gives an indication of the depths of his feelings about the quantum theory. He wanted to destroy it. Nonetheless, Einstein nominated Heisenberg and Schrödinger for the Nobel Prize in 1928 and again in 1931. In 1931, his nomination proposal said of quantum mechanics, "I am convinced that this theory undoubtedly contains a part of the ultimate truth."

After this encounter with Bohr, Einstein's attitude toward the quantum theory seems to have changed. He apparently accepted its logical consistency but refused to believe that it was any more than "a part of the ultimate truth." This attitude resulted in the publication in 1935 of what has become the most enduring residue of the Einstein-Bohr debates, Einstein's paper written with two young Princeton associates, Boris Podolsky and Nathan Rosen. (Einstein had emigrated to the United States in the summer of 1933.) The paper, which has the title "Can Quantum-Mechanical Description of Physical Reality Be Considered Complete?" was, unlike most of Einstein's previous papers, written in English and published in the *Physical Review*, the professional journal of the American Physical Society. Once again, it deals with an imaginary experiment, but like so many other works of the imagination this one has surpassed the intentions of its creators. In the first place, it is unlikely that they thought that the experiment they were proposing, or any facsimile, would ever be carried out. This, thanks largely to the work inspired by Bell, has now happened. Indeed, our physics journals are now resplendent with new and ever more ingenious versions of the Einstein-Podolsky-Rosen experiment, along with increasingly accurate experimental results. In the second place, and this is also an aftermath of Bell's work, the Einstein-Podolsky-Rosen experiment has made its way into much of the popular folklore about the quantum theory. (It is usually referred to in the literature, familiarly, as the EPR experiment.) This has much to do with an aspect of the experiment that was not even mentioned in the original paper, namely its use of what Einstein would later call "spooky actions at a distance" (*spukhafte Fernwirkung*, in the German original). This concerns influences of one part of a system on another even when the two parts are widely separated in space. I will come back to this matter shortly.

The EPR experiment has to do with the kind of inferential knowledge we make use of all the time correlating some event taking place in front of us with some event at a distant place beyond our ken because we know some relationship between the objects involved. I have already mentioned Bell's fetching example of Bertlmann's socks. Here is another equally fetching example provided by Bell. "Suppose," he writes, "I take from my pocket a coin and, without looking at it, split it down the middle so that the head and tail are separated. Suppose then, still without anyone looking, the two different pieces are pocketed by two different people who go on two different journeys. The first to look, finding that he has a head or tail, will know immediately what the other will subsequently find." In the language of Einstein, Podolsky, and Rosen, the observation by the first person that he has, say, a tail confers "reality" on the head that the second observer will inevitably find. In the somewhat formalistic language of the EPR paper, the authors propose the following definition (the italics are theirs): "*If, without in any way disturbing a system we can predict with certainty . . . the value of a physical quantity* [the head, in the example above] *then there exists an element of physical reality corresponding to this physical quantity*." According to this definition, the observation of tail makes the head *real* even though it is not directly observed.

Einstein, Podolsky, and Rosen went on to apply the same logic to quantum mechanics. For this purpose, they supposed that two particles have approached each other from a great distance, interacted, and then separated once again to a great distance. By an ingenious arrangement they set things up so that if measurement of the position of one particle is done, this immediately tells them the position of the other, far distant particle. This then, according to their definition, confers "reality" on the position of the second particle. Likewise by modifying their setup they can measure the momentum of the first particle in such a way that this tells them the momentum of the second particle. This confers "reality" on the second particle's momentum as well—that is to say, both the position and momentum of the second particle are given "reality." But quantum mechanics tells us that no quantum-mechanical description is possible in which both position and momentum are precisely specified. Hence in this arrangement it looks as if we have conferred "reality" on something that quantum mechanics cannot describe. Thus, their answer to the question in the title of their paper "Can Quantum-Mechanical Description of Reality be Considered Complete?" was a clearcut no.

The EPR paper took Bohr completely by surprise. Léon Rosenfeld recalled later that "this onslaught came down upon us as a bolt from the blue. Its effect on Bohr was remarkable." They were in the middle of exploring some puzzles that had arisen in the applications of the quantum theory to electromagnetism. Rosenfeld goes on, "A new worry could not come at a less propitious time. Yet, as soon as Bohr had heard my report of Einstein's argument, everything else was abandoned: we had to clear up such a misunderstanding at once. We should reply by taking up the same example and showing the right way to speak about it. In great excitement, Bohr immediately started dictating to me the outline of such a reply. Very soon, however, he became hesitant: 'No, this won't do, we must try all over again. . . . we must make it quite clear. . . . ' So it went on for a while, with growing wonder at the unexpected subtlety of the argument. Now and then, he would turn to me: 'What *can* they mean? Do *you* understand it?' There would follow some inconclusive exegesis. Clearly, we were further from the mark than we first thought.

Eventually, he broke off with the familiar remark that he must sleep on it.

"The next morning he at once took up the dictation again, and I was struck by a change in the tone of the sentences: there was no trace in them of the previous day's sharp expressions of dissent. As I pointed out to him he seemed to take a milder view of the case, he smiled. 'That's a sign,' he said, 'that we are beginning to understand the problem.' And indeed, the real work now began in earnest; day after day, week after week, the whole argument was patiently scrutinized with the help of simpler and more transparent examples. Einstein's problem was reshaped and its solution reformulated with such precision and clarity that the weakness in the critics' reasoning became evident, and their whole argumentation, for all its false brilliance, fell to pieces. 'They do it "smartly," ' Bohr commented, 'but what counts is to do it right.' " The whole process took six weeks, Rosenfeld recalls, a kind of speed record for Bohr.

Some indication of Bohr's eagerness to get his message out was the fact that prior to the publication of his rebuttal of the EPR paper in the *Physical Review*, he sent a letter to the British journal *Nature* announcing his forthcoming paper and assuring people that everything was really all right. Despite Rosenfeld's—and implicitly Bohr's—characterization of the EPR paper, there is nothing really "wrong" with it in the sense of a mistake by its authors in either the physics or the mathematics, of the kind that Einstein had done when he tangled with Bohr in 1930. If one could construct an EPR apparatus, there is no reason to think that it would not do just what its inventors said it would—namely, one could use it first to measure implicitly the exact position of a particle at a distant place. One could then modify the apparatus to measure the exact momentum of the distant particle. What the EPR paper does not state clearly is that these measurements are separate and *distinct*, and involve different modifications of the apparatus. In no single measurement can one measure *both* the position and momentum with arbitrary precision. If one uses the EPR setup to measure the position, then the momentum becomes completely uncertain, and vice versa. In one insists on using the EPR language one can say that one measurement has conferred "reality" on the position and the other has conferred "reality" on the momentum, but no experiment can confer "reality" on both if one believes in the correctness of quantum mechanics.

For Einstein, this was what was intolerable about the theory. He seemed to feel that once reality was conferred on something, this reality should be an enduring aspect of that thing, not to be removed because the experimenter had chosen to measure something different. For Bohr, on the other hand, and this is the burden of his response to Einstein, all of this was just another illustration of the principle of complementarity. A particle in *reality* has *neither* a position nor momentum. It has only the *potential* to manifest these complementary properties when confronted by suitable experimental apparatus. For Einstein this was an incentive to look for a deeper theory. The EPR paper ends, "While we have thus shown that the wave function does not provide a complete description of the physical reality, we left open the question of whether or not such a description exists. We believe, however, that such a theory is possible." For Bohr the matter had been settled. The deeper theory was quantum mechanics.

It is clear that these views were irreconcilable. While Bohr continued to write about these matters after this exchange, the EPR article was the last, as far as I know, that Einstein ever published in a technical journal about the quantum theory. But he thought about it incessantly; his correspondence with Max Bohr reflects that. Something of his state of mind can be seen from a letter he wrote to Born on September 7, 1944. He ends it by writing, "We have become Antipodean in our scientific expectations. You believe in the God who plays dice, and I in complete law and order in a world which objectively exists, and which I, in a wildly speculative way, am trying to capture. I firmly *believe*, but I hope that someone will discover a more realistic way, or rather a more tangible basis than it has been my lot to find. Even the great initial success of the quantum theory does not make me believe in the fundamental dice-game, although I am well aware that our younger colleagues interpret this as a consequence of senility. No doubt the day will come when we will see whose instinctive attitude was the correct one."

While it is unlikely that Einstein's younger colleagues went so far as to characterize him as being "senile," it is certainly true that few of them paid much attention to the issues raised by the Einstein-Podolsky-Rosen paper. This had largely to do with the ever-increasing practical success of the theory. By the 1930s Linus Pauling, and others, had explained, chemical bonding using quantum mechanics. About the same time, Heisenberg, Fermi, and others had shown that the theory, which had originally been invented to deal with atomic phenomena, had validity even at nuclear dimensions, which are some one hundred thousand times smaller than atoms. Dirac had made a marriage of the quantum theory and relativity in the late 1920s. Its offspring was the prediction of antimatter, which was confirmed in 1932 when the American physicist Carl Anderson found positrons—antielectrons—in cosmic rays. It is true that there were some puzzles to clear up in the applications of quantum theory to electricity and magnetism, but the optimism of the day is well reflected in Dirac's aphorism that, even then, quantum theory explained "most of physics and all of chemistry." That a few elderly physicists were squabbling about the foundations of the theory seemed all but irrelevant.

This attitude was reflected in quantum-mechanics textbooks, not only in the ones written before the war, but even until fairly recently. In the early 1980s, before the impact of the work initiated by Bell had found its way into textbooks, I made a survey of seventeen standard textbooks on the quantum theory. I discovered that only one of them, David Bohm's *Quantum Theory*, published in 1951, made any reference at all to the paper of Einstein, Podolsky, and Rosen. (I will have more to say about Bohm's book shortly.) Even Dirac's great text, *Quantum Mechanics*, which Einstein had called the "most logically perfect presentation" of the theory, has no mention of the paper nor of any of the issues it raises. Bell told me that in the 1950s a colleague of Dirac's asked him why none of this was discussed in his book. Dirac answered, in his usual elliptic way, "I think it is a good book, except for the absence of several introductory chapters." These were, presumably, the chapters in which Dirac might have discussed the foundations of the theory. He seems to have decided that these questions were too far removed from resolution and of too little relevance to application to merit inclusion in his book.

John Bell began reading popular books on the quantum theory while he was still in high school. He began taking courses in it during his last two years at the Queen's University in Belfast. This was in the late 1940s, and none of the texts discussed the more philosophical aspects of the theory. Bell remembers, with some distaste, a course he took on atomic spectra—bread and butter quantum mechanics. "It went into what I thought at the time was a lot of unnecessary detail," he said. "All the atoms in the periodic table. I think that atomic spectra were something of a speciality at Belfast because they were interested in the physics of the upper atmosphere. We learned a lot of stuff about the Bohr atom. That might make you wonder what was happening while the electron was jumping from one orbit to another as the atom is radiating. But still—there are rules. You learn about the periodic table of elements—all the practical aspects of the theory. Then the puzzles start."

Bell recalls being particularly perplexed by the Heisenberg uncertainty principle. "It looked as if you could take this size and then the position is well defined, or that size and then the momentum is well defined. It sounded as if you were just free to make it what you wished. It was only slowly that I realized that it's not a question of what you wish. It's really a question of what apparatus has produced this situation. But for me it was a bit of a fight to get through to that. It was not very clearly set out in the books and courses that were available to me. I remember arguing with one of my professors, a Doctor Sloane, about that. I was getting very heated and accusing him, more or less, of dishonesty. He was getting heated too and said, 'You're going too far.' But I was very engaged and angry that we couldn't get all that clear." I can imagine Bell at nineteen, his hair probably redder than it is now, his Irish temper flaring, because Dr. Sloan could not explain the uncertainty principle clearly enough.

It was at this time that Bell began thinking about the problem that has bothered him ever since: where does the quantum world leave off and the classical world begin? In our everyday lives we do not directly experience any of the bizarre effects described by the quantum theory. Does that mean that the quantum theory does not apply to us, but only to our individual atoms? How many atoms does it take before we have a system that is large enough so that it becomes classical? In 1935 Schrödinger invented a paradox that shows the conundrums one can get into if one supposes that quantum mechanics applies to systems as big as ourselves or, in his case, as big as cats. He imagined a dastardly arrangement in which, as he put it, "a cat is penned up in a steel chamber, along with the following diabolical device (which must be secured against direct interference by the cat): in a Geiger counter there is a tiny bit of radioactive substance, *so* small that *perhaps* in the course of one hour one of the atoms decays, but also with equal probability perhaps none; if it happens, the counter tube discharges and through a relay releases a hammer which shatters a small flask of hydrocyanic acid. If one has left this entire system to itself for an hour, one would say that the cat still lives *if* meanwhile no atom has decayed. The first atom decay would have poisoned it." If one insisted on using quantum mechanics to describe this entire system, then one would have to say that prior to observing the cat, by opening the steel box, it would be neither dead or alive but in some weird quantum-mechanical mixture of life and death, which seems like an absurdity.

I asked Bell if he had, as a student, taken up reading people like Bohr, especially about questions like these. He said that he had, and went on, "I disagree with a lot of what Bohr said. But

I think he said some very important things which are absolutely right and essential. One of the vital things that he always insisted on is that the apparatus [the measuring instrument] is classical. For him there was no way of changing that. There must be things we can speak of in a classical way, and for him that was the apparatus. For him it was inconceivable that you could extend the quantum formalism to include the apparatus. [If you did you would confront all sorts of Schrödinger cat–like paradoxes.]

"It is very strange in Bohr that, as far as I can see, you don't find any discussion of where the division between his classical apparatus and the quantum system occurs. [How many atoms does it take to construct a classical apparatus?] Mostly you will find that there are parables about things like a walking stick—if you hold it closely it is part of you, and if you hold it loosely it is part of the outside world. He seems to have been extraordinarily insensitive to the fact that we have this beautiful mathematics, and we don't know which part of the world it should be applied to.

"Bohr seemed to think that he had solved this question. I could not find his solution in his writings. But there was no doubt that he was convinced that he had solved the problem and, in so doing, had not only contributed to atomic physics, but to epistemology, to philosophy, to humanity in general. And there are astonishing passages in his writings in which he is sort of patronizing to the ancient Far Eastern philosophers, almost saying that he had solved the problems that had defeated them. It's an extraordinary thing for me—the character of Bohr—absolutely puzzling. I like to speak of two Bohrs: one is a very pragmatic fellow who insists that the apparatus is classical, and the other is a very arrogant, pontificating man who makes enormous claims for what he has done."

I asked Bell where he thought disciples of Bohr such as Heisenberg and Pauli stood on these questions. After a little thought he answered, "In the beginning Heisenberg and Pauli felt very close to Bohr. Those three were the Copenhagen trio, the Three

Musketeers of the Copenhagen interpretation. In later years, Pauli seems to have decided that Bohr himself was not a complete supporter of the Copenhagen interpretation. He reproached Bohr along the following lines: Bohr insisted that there was this division between the quantum-mechanical system and the classical apparatus. He explicitly repudiated the idea that the human mind was somehow an important element in quantum mechanics—that is, that the division was between the interior world [mind] and the outer world [matter]. But Pauli was attracted to that idea, and at the end of his life [Pauli died in 1958 at the age of 58] he became increasingly religious. He felt that it was wrong to separate science from religion; it was wrong to separate psychology from physics. He felt that the *real* Copenhagen interpretation did insist that the mind was something that you could not avoid referring to in formulating quantum mechanics. Pauli thought, as far as I can judge, that the division between system and apparatus was ultimately between mind and matter.

"It is perfectly obscure to me *what* Heisenberg thought. As I said, Bohr thought it was between big objects and little ones and seems to have been remarkably insensitive to the need to make that distinction more precise. For me, it was a big risk that I would get hung up on these questions once I learned about them. When I was not quite twenty-one, I rather deliberately walked away from them, and toward accelerator physics. I had the feeling then that getting involved in these questions so early might be a hole I wouldn't get out of."

In the early 1950s, while Bell was working on accelerators, there were two important advances in the interpretation of quantum mechanics, both made by the American physicist David Bohm. Just as there were, it seems, two Bohrs, there were apparently two Bohms: a 1951 Bohm and a 1952 Bohm. The 1951 Bohm, who was then in Princeton, published an influential textbook entitled *Quantum Theory*. It is the only serious textbook on the quantum theory that I have ever seen in which there are more words than equations. Unlike all its predecessors and most of its successors, Bohm goes into great detail about the interpretation of the theory. All of this is done from the orthodox Bohr point of view. He presents a novel version of the Einstein, Podolsky, and Rosen experiment, one that is much easier for someone trying to learn the subject to visualize. It has become the basis of most of the modern discussion. It is this version, especially following the work of Bell, that has lent itself to actual laboratory experiments. The purpose of Bohm's discussion of the EPR experiment is to argue, along the lines of Bohr's refutation, that the experiment, properly interpreted, is simply another example of the principle of complementarity. Toward the end of the book Bohm presents an extensive discussion of the "hidden variable" question: This is important for understanding Bell's work, so I will say a few words about it here and come back to it later.

When Einstein spoke of "completing" the quantum theory, it was never entirely clear what he meant. It probably could not have been made completely clear unless he had succeeded in carrying out the program; whatever that turned out to be. One possibility for what he meant, and there is support for this in his writing, would be "hidden variables." The paradigmatic example of the introduction of hidden variables into physics was the use of the atomic theory of matter by such nineteenth-century physicists as James Clerk Maxwell and Ludwig Boltzmann to explain the laws of thermodynamics, such as the conservation of energy, which had been discovered earlier in the century. It would have been perfectly consistent to take these laws at their face value without searching for a deeper meaning. However, people like Maxwell and Boltzmann argued that these laws could be "explained" if one supposed that, says, a heated gas consisted of chaotically moving atoms and that heat, for example, was nothing but a manifestation of the random motion of these atoms. To us, this seems almost obvious, but in the nineteenth century no one had ever seen an atom. One was proposing to explain observed macroscopic regularities by introducing another, apparently hidden domain of unobserved and possibly unobservable microscopic phenomena. Many physicists, even some of Planck's caliber, resisted the atomic hypothesis as an unnecessary complication.

Two of Einstein's 1905 papers deal with just this question. As he wrote to his friend Habicht, in the letter I quoted, he used what he called the "molecular theory of heat" to account for the disordered motion of small particles suspended in liquids, the so-called Brownian motion. These small particles were visible, at least in a microscope, whereas the molecules were not. In this sense the molecules represented the hidden, or uncontrollable, microscope variables that governed the destinies of the observed macroscopic particles. Could there not be something similar going on in the quantum theory? Indeed, in a 1948 article Einstein put the matter in just these terms. He wrote, "Assuming the success of efforts to accomplish a complete physical description, the statistical quantum theory would, within the framework of future physics, take an approximately analogous position to the statistical mechanics within the framework of classical mechanics. I am rather firmly convinced that the development of theoretical physics will be of this type; but the path will be lengthy and difficult."

When quantum mechanics was first invented, other members of the older generation (apart from Einstein), including de Broglie and Schrödinger, had misgivings about the developing interpretation of the theory. In late 1926, as Heisenberg afterward recalled, Schrödinger paid a visit to Copenhagen. He was still trying to defend the idea that the Schrödinger waves were oscillations in actual space. Bohr was relentless. The discussion went on night and day until finally Schrödinger, exhausted, retreated to his bed. Bohr felt that this was no reason to end the discussions and followed him into the bedroom. Heisenberg reports, "[The] phrase 'But Schrödinger, you must at least admit that . . . ' could be heard again and again." Finally, in desperation, Schrödinger said, "If we are going to stick to this damned quantum jumping, then I regret that I ever had anything to do with quantum theory." The irrepressible George Gamow, who was in Copenhagen a few years later, drew a cartoon that showed a similar scene with the Russian physicist Lev Landau bound and gagged, with Bohr hovering over him and saying, "May I get a word in?"

By the early 1930s, with the exception of Einstein and Schrödinger, most of this opposition had died down. In 1932, the remarkable Hungarian-born mathematician John von Neumann published a book entitled *Mathematische Grundlagen der Quantenmechanik (Mathematical Foundations of the Quantum Theory)*. He presented the quantum theory as if it were a branch of pure mathematics, like Euclidean geometry, derivable from a set of formal axioms. It was an immensely influential book, and greatly clarified the mathematical foundations of the theory. In the book he presented what purported to be a formal proof that no hidden-variable theory could reproduce the results of quantum mechanics.

On its face, this argument would seem to have ruled out any attempt along the lines that Einstein apparently had in mind. I am not aware that Einstein ever commented on von Neumann's proof. Perhaps he did not know about it or, and I think this is more likely, he had a certain mistrust when it came to mathematical proofs of things being impossible in physics. In any event, Bohm presented a version of this proof in his text and drew the conclusion, in agreement with von Neumann, that hidden-variable theories were impossible. That was the Princeton Bohm.

Bohm had come to Princeton in 1947 from Berkeley. He had been a student of Robert Oppenheimer's, and after Oppenheimer moved to Princeton Bohm followed him to the university. During the war he had wanted to go to Los Alamos but had been turned down for security reasons. In 1949, just before his text came out, he became the subject of an investigation by the House Un-American Activities Committee before which he appeared in the spring of 1949. He pleaded the Fifth Amendment and was indicted for contempt of Congress. Despite the fact that he was acquitted, the president of Princeton refused to reappoint him, and he was unable to find another job in this country. However, helped by a strong letter of recommendation from

Einstein, in 1951 he was offered a job at the University of São Paulo in Brazil. He had been there only a few weeks when he was summoned by the American Consul, who removed his passport. He was informed that the passport would be stamped "VALID ONLY FOR RETURN TO THE UNITED STATES." The Consul, apparently, feared that Bohm would visit the Soviet Union.

Bohm's concern was that he would become scientifically isolated in Brazil. Hence he applied for and obtained Brazilian citizenship so that he could travel. In 1955 he left Brazil and spent two years in Israel before settling in England, where he recently retired from Birkbeck College of the University of London. Thus this country lost the services of one of the best American physicists of the postwar generation.

When he was in Brazil, Bohm published two papers, one of which ends with an acknowledgment thanking Einstein for "several interesting and stimulating discussions." In the course of these discussions Einstein seems to have persuaded Bohm to look again at the interpretation of quantum mechanics. Murray Gell-Mann, who saw Bohm frequently at this time, remembers Bohm's crediting his conversion to discussions with Einstein. Be that as it may, the content of Bohm's Brazil papers was just to exhibit a specific model of exactly the sort of hidden-variable theory that the Princeton Bohm had "proven" to be impossible. Clearly something very murky was going on. I will return to these matters shortly.

Before I do so, I would like to discuss Bohm's version of the EPR experiment, the one given in his textbook and which is used as the basis of the modern treatments of the subjects. To do this, I have to introduce the notion of "spin," a remarkable quantum-mechanical attribute that elementary particles can have and a fascinating subject in its own right. Spin is a form of angular momentum. First, then, what is an *angular* momentum?

When, say, a pair of stars—double stars—orbit around each other, the ordinary momentum of the system is zero. But their circular motion is evident. Physicists introduced a second kind of momentum, angular momentum, to characterize this kind of motion. The angular momentum of the double stars is called "orbital" angular momentum, and it has the property that it vanishes when the orbiting object is brought to rest.

The "old" quantum theory used orbital angular momentum as one of the characterizations of the different Bohr orbits of the electrons orbiting around an atomic nucleus. By 1925 it became apparent that not all the atomic spectral lines could be accounted for if the only characterization of the orbits was the orbital angular momentum. Hence the notion arose that the electron should be allotted an additional angular momentum, which became known as its spin. This angular momentum, which has no classical counterpart, persists even after, say, the electron is brought to rest. The image, which does not really do justice to the quantum-mechanical abstraction, is of the spinning electron as a tiny top whose spinning persists even if the electron is brought to rest.

The concept of spin was invented in 1925 by the young American theoretical physicist Ralph Kronig. Unfortunately for him, he was talked out of publishing it by Pauli, who at first thought that the idea was nonsense. He used to refer to it as "*Irrlehre*" ("heresy"). So it had to be independently reinvented by two equally young Dutch physicists, Samuel Goudsmit and George Uhlenbeck, also in 1925. They were at Leiden and were working under the general guidance of Einstein's friend Paul Ehrenfest. Ehrenfest advised the young men to write up their work so that it could be shown to the great senior theoretical physicist at Leiden, Hendrik Antoon Lorentz. Uhlenbeck recalled many years later that in a few days they got back a long manuscript, filled with equations, from Lorentz. The burden of the manuscript was also that spin was nonsense. Lorentz took the picture of the spinning electron quite literally. He thought of it as a sort of classical particle with a certain extension and argued that, if it were spinning as fast as Goudsmit and Uhlenbeck claimed, its surface would be spinning faster than the speed of light, which would have violated the basic principle of the theory of relativity that nothing can move faster than light in a vacuum. A modern quantum theorist thinks of the electron, in a certain sense, as a point particle, so this objection is irrelevant.

At the time, Goudsmit and Uhlenbeck were crushed. Uhlenbeck wrote, "Goudsmit and myself both felt that it might be better for the present not to publish anything; but when we said this to Ehrenfest, he answered, '*Ich habe Ihren Brief schon längst abgesandt. Sie sind beide jung genug um sich eine Dummheit leisten zu können!*' ['I sent your paper off long ago. You're both young enough that you can afford a stupidity!']"

A classical angular momentum can have any value at all. Spin, however, can have only the values 0, 1/2, 1, 3/2, 2, 5/2 . . . and so on (in suitable units). A classical angular momentum can point in any direction, and there are no limitations on the accuracy with which that direction can be measured. With spin there are limitations, and these are given by a variation of the Heisenberg uncertainty principle.

Let us take spin 1/2 as the simplest illustration. If we pick an axis that defines a direction and measure (I will explain how this is done shortly) the direction of the spin with respect to that axis, then we will discover that the spin can point only along the axis or in the opposite direction. In the first instance we say that the spin is "up," and in the second we say that the spin is "down." Having chosen an axis, and having made our measurement, we can then change axes and make a new set of measurements. Once again we will find that the spin points up or down with respect to the new axis.

This is reminiscent of the situation with position and momentum. As we have seen, at least in the usual interpretation a quantum-mechanical particle has neither a position nor a momentum. It has the potential for exhibiting a position, say, when confronted by a measuring instrument designed to measure positions. In the same sense a spin has no direction, but it has the potential for pointing "up" or "down" in a measurement designed to measure directions. Remember that Bohr warned that one could get *schwindlig* (dizzy) thinking about quantum mechanics. Spin may be a case in point. The fact that spin can point only in a restricted number of directions with respect to a given axis (two for spin 1/2, 3 for spin 1, and so on) was given the name "space quantization," although it is not space that is quantized but rather the directions in which the spin can point.

In actual fact, space quantization was discovered experimentally before the invention of spin. In 1922 the German physicists Otto Stern and Walter Gerlach discovered space quantization in an experiment involving silver atoms. They heated silver atoms in a furnace and extracted a beam of these atoms, which they allowed to pass through a strong magnetic field. The direction of this field is what defines the spatial axis of which we spoke before. Not knowing anything about spin, they assumed that the silver atoms, after their encounter with the magnet, would have had their trajectories bent in essentially arbitrary and random directions. To test this they allowed the atoms to deposit themselves on a plate of glass. They expected to see a kind of smear of atoms. Instead, the atoms deposited themselves into two fine lines separated by a fraction of a millimeter.

Nowadays, having absorbed the lessons of the quantum theory, we would say that this was a perfect demonstration that silver atoms have spin 1/2. The atoms could move only "up" or "down" in the magnetic field, so half of them moved up and half moved down—hence the two fine lines. However, to physicists at the time, this result was absolutely astounding. The late I. I. Rabi, who was entering physics at about the time of the Stern-Gerlach experiment and who a few years later went to Germany to work with Stern, once said to me, "I thought the old quantum theory was stupid. I thought one might be able to invent another model of the atom which had the same properties. But you can't get around the Stern-Gerlach experiment. You are really confronted with something quite new. It goes on in space, and no clever classical mechanism would do—would explain it."

In the Bohm version of the Einstein-Podolsky-Rosen experiment, we imagine that we have at our disposal two Stern-Gerlach-like magnets located at different places, far enough apart that they have no contact with each other. We may imagine that near one of the magnets we have a supply of molecules, the analogue of the silver atoms in the original Stern-Gerlach experiment. These molecules, we suppose, are each composed of two atoms, and each of these atoms has spin 1/2. It is possible to arrange things so that, loosely speaking, the spins of the atoms in these molecules point in opposite directions, so that the net molecular spin is zero. The resultant molecule is spinless.

Now imagine, with Bohm, that the molecules spontaneously split up—a kind of radioactivity—into their constituent atomic components. When any given molecule splits up, its two atoms fly off in opposite directions into the waiting Stern-Gerlach magnets. An observer at one of the magnets records whether or not the atom that enters his magnet has spin up or down. If the answer is spin up, then that observer can be one-hundred-percent certain that the observer at the other magnet would record a spin down. This reflects the correlations of the atomic spins in the original molecule. If we now use the criterion of EPR we would say that this experiment has conferred "reality" on the direction of the spin of the second atom.

Now we may imagine rotating both magnets by, say, ninety degrees and doing the experiment again. According to EPR this will confer "reality" to a different direction of the spin. But the Heisenberg principle, in this case, tells us that no experiment can measure the components of spin in more than one direction. Hence EPR would conclude that quantum mechanics does not offer a complete description of reality, while Bohr would argue that this setup is just another illustration of the principle of complementarity. In this case, the complementary aspects are the components of the spins in more than one direction.

All of this is clearly spelled out in Chapter 22 of Bohm's book. There are at least two virtues to the Bohm version of the EPR experiment as opposed to the original. In the first place, one can begin to imagine how one might do a *real* EPR experiment, since measuring spins with Stern-Gerlach-like experimental techniques has become, over the years, common practice for experimenters. I will come back to this shortly. The second virtue is that this setup makes it graphically clear just how bizarre these distant quantum-mechanical correlations are altogether.

To make things as graphic as possible, let us suppose that one Stern-Gerlach magnet is in Princeton and the other one is on the moon, or even farther away. And again let us suppose that there is no contact between the two observers. Let us suppose that before departing for the moon the observers agree to set their magnets so that the field directions are at right angles to each other. Each observer makes a tape that shows whether the spin is up or down when each atom arrives at its respective magnet. Afterward the tapes are compared event by event to see if there are any correlations between the spin measurements made at the different magnets.

If the magnets are actually set at right angles, quantum mechanics predicts that there are no correlations. A spin up at one magnet is just as likely to be associated with a spin up at the other as with a spin down. But suppose that while the atoms are traveling through space, one of the observers changes his mind and does not rotate his magnet, so the magnets remain parallel. After the experiments are done and the tapes are examined, what the observers will now find is *perfect* correlation between the events. Every time a spin up is recorded by one observer a spin down will have been recorded by the other. These correlations manifest themselves according to the rules of quantum mechanics even though there is no contact, until long after the fact, between the two observers. This is surely what Einstein had in mind when, in connection with the quantum theory, he spoke of "*spukhafte Fernwirkungen*"—spooky actions at a distance. Someone not trained in the quantum theory looking at these experimental results would surely conclude that the atoms must be carrying with them some kind of program—some kind of "hidden variable"—which instructs them in some sort of causal fashion on how to respond when confronted with the magnets in their various settings.

Bell is often called upon to lecture on these subjects to groups of people, such as secondary school teachers, who have had no special scientific training. To help them understand the oddness of these distant correlations, he compares them to the situation of those rare pairs of identical human twins who have grown up in separate environments with no knowledge of each other. The twins correspond to the two atoms, which are "born" in the same decay, and then lose contact with each other. If one followed the behavior of one of the twins (atoms) with no reference to the other, there would be no suggestion of anything in that behavior that that person was a twin. In the same way, an individual atom arriving at one of the magnets will move up or down in an apparently random way. It is only when the tapes are compared for the two atoms that the correlations emerge. Likewise, it is only when twins finally meet, if they ever do, that remarkable correlations in their behavior appear.

Bell is fond of one specific example he turned up while perusing the annals of an entity called The Institute for the Study of Twins. He noted with some astonishment that they had "found two people, identical twins, who had been separated as babies. There was an amazing list of coincidences between them. They both bit their nails. They both smoked the same brand of cigarette. They had both bought the same model of car. They bought it in the same color. They went to the same resort in Florida for their holidays with their families. They had married on the same day. They both had dogs, and the dogs had the same names. In my lecture I show a picture of them."

In the case of the twins, we have what we think of as a "scientific" explanation of these coincidences. These twins have identical genes. Genes, we think, are an important factor in determining behavior. Hence this pair of identical twins is a beautiful illustration of genetic determinism. According to the conventional quantum theorist, it is here that the analogy between the twins and the atoms breaks down. Such a theorist would argue that nothing "explains" the EPR correlations; they are simply a fact of life. Indeed, the theorist would say more. As Bell puts it, "If this theorist was one of the more dogmatic members of the Copenhagen school, he would say, 'It's *naive* to demand an explanation. I give you the rule for the correlation and that's enough.' " This is the attitude that is brilliantly defended in Bohm's textbook. It is also the essential implication of von Neumann's theorem.

Bell first heard about the theorem when he was a student at the university in Belfast. "I read about it," he told me, "in a book by Max Born called *Natural Philosophy of Cause and Chance*. I was very impressed that somebody—von Neumann—had actually *proved* that you couldn't interpret quantum mechanics as some sort of statistical mechanics." In statistical mechanics, remember, the atoms act as "hidden variables" whose causal behavior accounts for the laws of thermodynamics. However, as I have mentioned, in 1952 Bohm (who was then at the University of São Paulo) published two papers in the *Physical Review* entitled "A Suggested Interpretation of the Quantum Theory in Terms of 'Hidden' Variables"—exactly what von Neumann had "proved" was impossible, and which Bohm had echoed in his 1951 text.

In the introduction to the first of these papers Bohm explains their intent. He writes, "Most physicists have felt that objections such as those raised by Einstein [to the quantum theory] are not relevant, first, because the present form of the quantum theory with its usual probability interpretation is in excellent agreement with an extremely wide range of experiments, at least in the domain of distances larger than 10^{-13} centimeters, and secondly, because no consistent alternative interpretations have as yet been suggested. The purpose of this paper . . . is, however, to suggest just such an alternative interpretation. In contrast to the usual interpretation, this alternative interpretation permits us to conceive of each individual system as being in a precisely definable state, whose changes with time are determined by definite laws, analagous to (but not identical with) the classical equations of motion. Quantum mechanical probabilities are regarded (like their counterparts in classical statistical mechanics) as only a practical necessity and not as a manifestation of an inherent lack of complete determination in the properties of matter at the quantum level." In other words, Bohm claimed to have reduced quantum mechanics to a new form of classical physics free of probabilities, indeterminism, and all the rest.

(Since Bohm wrote this, the validity of quantum mechanics has been shown to extend down to distances much shorter than this. The length 10^{-13} centimeters is known as the "Fermi," after Enrico Fermi. It is roughly the size of the smallest nucleus, the nucleus of hydrogen, known as the proton. The *interior* of the proton has now been explored and is well described by the quantum mechanics of the quarks.)

Bell was smitten by Bohm's papers. When he described his reaction, even after nearly forty years, he was still excited. "I couldn't read von Neumann's book," he told me, "because it was only available in German, and I couldn't read German. Then Bohm's papers came out in 1952. I was enormously impressed with them. I saw that von Neumann must have been just wrong. At that time, I discussed these things with a colleague of mine at Harwell, Franz Mandl. Franz was of German origin, so he told me something of what von Neumann was saying. I already felt that I saw what von Neumann's unreasonable axiom was." It turned out that von Neumann had made what appeared to be an innocuous technical assumption for which there was no justification. No one had paid any attention to this until Bohm produced his model. Of course, Bohm was aware that his model contradicted von Neumann's theorem. At the end of his second paper he discusses this. Having read this discussion several times, I think it is fair to say that it is, to put it charitably, obscure. The matter was clarified only a few years later by Bell, who isolated the specific unreasonable mathematical assumption von Neumann had made.

I myself was studying quantum mechanics as a student at Harvard in 1952 when Bohm's papers were published. I can recall mixed—indeed, largely negative—reactions to them. Part of the problem was that they did not seem to contain any new physics. Bohm was perfectly aware of this. After all, the idea of his paper was to demonstrate that quantum mechanics could be reinterpreted as a deterministic theory with hidden variables. The validity of quantum mechanics was assumed. As I have mentioned, Bohm felt that quantum mechanics had been tested only down to distances as small as 10^{-13} centimeters. He thought that quantum mechanics might break down at smaller distances and that new physics would reveal itself. Possibly, he felt, this new way of looking at quantum mechanics would play a role in this new physics. But, as I have also mentioned, we have now explored distances much smaller than this, with no sign that the theory is going to break down.

One might naively have thought that Einstein, with whom Bohm had discussed these matters would have been enthusiastic about Bohm's new way of looking at things. Not at all. He ends a letter written to Max Born on May 12, 1952, with the following comment: "Have you [Born] noticed that Bohm believes (as de Broglie did, by the way, 25 years ago) that he is able to interpret the quantum theory in deterministic terms? That way seems too cheap to me. But you, of course, can judge this better than I."

I was so surprised when I came across this reaction of Einstein's to what, offhand, I would have thought was the kind of development he would have liked, that I asked Bell what he thought it meant. Before replying directly, Bell said with some heat that among the books he would like to write—"I would like to write a half a dozen books, which means I won't write any"—would be one tracing the history of the hidden-variable question and especially the psychology behind people's peculiar reactions to it. "Why were people so intolerant of de Broglie's gropings and of Bohm?" he asked me. Without waiting for an answer, he went on, "For twenty-five years people were saying that hidden-variable theories were impossible. After Bohm did it, some of the same people said that now it was trivial. They did a fantastic somersault. First they convinced themselves, in all sorts of ways, that it couldn't be done. And then it becomes 'trivial.' I think Einstein," Bell went on, "thought that Bohm's model was too glib—too simple. I think he was looking for a much more profound rediscovery of quantum phenomena. The idea that you could just add a few variables and the whole thing [quantum mechanics] would remain unchanged apart from the interpretation, which was a kind of trivial addition to ordinary quantum mechanics, must have been a disappointment to him. I can understand that—to see that that is all you need to do to make a hidden-variable theory. I am sure that Einstein, and most other people, would have liked to have seen some big principle emerging, like the principle of relativity, or the principle of the conservation of energy. In Bohm's model one did not see anything like that."

When Bell went to Birmingham in 1953 to work with Peierls, he experienced firsthand this attitude on the part of physicists like Peierls, who had grown up with the conventional interpretation of quantum mechanics. All the students were asked to give a talk about what they had been doing. Bell gave Peierls a choice of two topics: the foundations of quantum theory or accelerators. Peierls made it quite clear that he preferred accelerators. To give an idea of how things have changed in recent years—owing in many ways to Bell's work—in 1979 Peierls published a book entitled *Surprises in Theoretical Physics* in which he discussed some of the things he had learned in a lifetime of doing theoretical physics that had surprised him. One of the surprises was the fact that von Neumann's theorem had not settled the matter of the hidden variables, which, he noted, had finally been clarified by Bell.

When the Bells take one of their rare leaves from CERN, they tend to go to California, to the Stanford Linear Accelerator. This may account for the fact that "Hawaiian" shirts seem to be one of Bell's sartorial preferences. Like nearly everyone else at CERN, both Bells dress very informally at work. Seeing Bell at CERN with a tie is a great rarity and usually signals some official function.

In any event, in late 1963 the Bells went to Stanford, where they had been invited to spend a year. As it happened, they arrived in the United States the day after President Kennedy was assassinated. "It was the worst possible moment to have come," Bell commented. Nevertheless, they were welcomed at the laboratory, where, as Bell remembers, "Mary was quickly integrated into the accelerator team at Stanford, and I into the theory group. Not long before I came over, I had once again begun considering the foundations of quantum mechanics, stimulated by some discussions with one of my colleagues, Josef Jauch. He, it turned out, was actually trying to strengthen von Neumann's infamous theorem. For me, that was like a red light to a bull. So I wanted to show that Jauch was wrong. We had gotten into some quite intense discussions. I thought that I had located the unreasonable assumption in Jauch's work. Being at Stanford isolated me, and gave me some time to think about quantum mechanics. My head was full of the argument of Jauch, and I decided that I would get all that down on paper by writing a review article on the general subject of hidden variables. [It did not appear until 1966, two years after Bell's theorem had been published.] In the course of writing that I became increasingly convinced that 'locality' was the center of the problem."

"Locality," in the sense that physicists now use the term, is something that came into physics quite recently, although its roots go back as far as the seventeenth century—the age of Newton. Prior to Newton the forces that physicists dealt with were the tangible forces of pushing and pulling. The object exerting the force was in direct contact with the subject of the force. Newton introduced the force of gravitation, which acted over long distances between objects that were not in any apparent physical contact. For example, the Earth, Newton claimed, acted on the Moon with the force of gravitation although the Earth and Moon do no appear to have any direct physical contact with each other.

It should be noted that prior to Newton, Descartes had also introduced a theory of gravitation. In his theory, space was filled with a medium called the "ether," and the forces of gravitation were transmitted through the ether in the same sort of way that water waves can transmit a force from one part of a body of water to another—by direct physical contact. When he was young, Newton also seemed to seek an "explanation" of gravitation in terms of an ether, but in his later years he took the position that the mathematical law of gravitation—the fact that the force of gravity falls off as the inverse square of the distance between two masses—did not need an explanation. This is the meaning I attach to his celebrated dictum that "the main Business of Natural

Philosophy is to argue from Phaenomena without feigning Hypotheses." If I may be pardoned the anachronism, I would call this the "Copenhagen interpretation" of gravitation.

In the nineteenth century the attention of physicists turned from gravitation to electricity and magnetism. At the end of the eighteenth century the French physicist Charles Augustin Coulomb discovered that electric charges attracted or repelled each other also by an inverse-square law. This created the science of electrostatics. Electrodynamics, the study of how electrical phenomena can vary in time and space, was created in the middle of the nineteenth century by Maxwell. One of Maxwell's accomplishments was to show that electricity and magnetism were related phenomena; a moving electric charge, for example, generates a magnetic field. His greatest contribution was providing the equations that relate electric and magnetic fields. He also showed that light was an electromagnetic phenomenon—a special kind of electric and magnetic field that oscillated in space and time.

Maxwell, and the rest of the physicists of his generation, believed that these oscillations took place in a material medium—the ether, again. The counterintuitive idea that electric and magnetic waves could propagate in empty space was not, as far as I know, considered at all. The first person in modern times to raise the question of the speed of this propagation was Galileo. In 1638 Galileo published his final scientific work, entitled *Discourses and Mathematical Demonstrations Concerning Two New Sciences*. The discourses are in the form of dialogues among three fictitious characters called Salviati, Simplicio, and Sagredo. (Sagredo probably represents Galileo himself.) One of the dialogues concerns the speed of light. Simplicio says,

> Everyday experience shows that the propagation of light is instantaneous; for when we see a piece of artillery fired, at a great distance, the flash reaches our eyes without lapse of time; but the sound reaches the ear only after a noticeable interval.

To this Sagredo-Galileo responds,

Well, Simplicio, the only thing I am able to infer from this familiar bit of experience is that sound, in reaching our ear, travels more slowly than light; it does not inform me whether the coming of light is instantaneous or whether, although extremely rapid, it still occupies time . . .

Sagredo proposes a method for deciding. He and an assistant will each have a lantern and will go to separate locations. Sagredo will then uncover his lantern, and when the assistant sees the flash, he will in turn uncover his own lantern. By measuring the time lapses involved one could in principle measure the speed of light. With modern electronics this experiment, or something like it, can be done, and the speed of light measured this way. But light is much too fast—the elasped times are much too small—for Galileo to have concluded in the seventeenth century anything except that the speed of light was very great. However, in 1674 the Danish astronomer Ole Roemer actually measured the speed of light using the moons of Jupiter, which are periodically eclipsed by the planet, as the distant lantern. The great distance to Jupiter makes this time lapse measurable. He found a value of 214,000 kilometers per second. This can be compared with the present value, obtained using the best modern techniques: 299,792.458 kilometers per second. Roemer was not that far off.

To the nineteenth-century physicists, the speed of light was no different from any other velocity. For example, for them it was quite conceivable that, given suitable propulsion, one might travel at the speed of light—or faster. This notion was completely altered after Einstein's special theory of relativity, one of his 1905 papers, became a generally accepted part of physics. (Incidentally, his theory of relativity was accepted much more rapidly than his ideas on the quantum theory.)

In the theory of relativity, the propagation of light is singled out. In the first place, the theory predicts that no material object can move faster than the speed of light. All the experiments ever done on rapidly moving particles (especially those in large accelerators, which move at speeds very close to that of light) bear this out. In the second place, the speed of light appears the same to all observers. To see how odd this is, suppose we have two observers, one at rest and the other moving in a spaceship at speeds close to that of light. Suppose the observer at rest has a flashlight, which he turns on. Both observers are then invited to measure the speed of propagation of the light emitted by the flashlight. The theory of relativity predicts that both observers will get exactly the same answer, even though the second observer is in rapid motion with respect to the first. This is possible only because space and time appear different to the two observers, and the distortions are just such as to maintain the absolute character of the speed of light.

The fact that no signal can propagate faster than the speed of light—no force and no influence—is what physicists call "locality." It means that to influence some event in the future, at some other place, there must be enough time allowed for the signal to propagate to that place. The notion of locality also gives us another view of the present instant. As I write this, I imagine the universe at exactly the same instant. I believe it exists. But I can never be quite sure. What reach me at the present instant are only signals from the past, propagating at the speed of light—or less. The speed of light is so fast that, for most practical purposes, I ignore this delay. I act as if these past events are really present.

Astronomers, on the other hand, confront the limitations imposed by locality all the time. When they say, for example, that a supernova has exploded in a distant galaxy, they speak as if that is a contemporary event. What is happening now is really only that a signal from the past has arrived at our telescopes now. By "now" that galaxy may have disappeared.

If we construct a theory, quantum-mechanical or otherwise, that is consistent with the theory of relativity, then it has this locality property built into it. Some attempts have been made to construct theories that have particles in them that *always* move faster than light, objects that the physicist Gerald Feinberg has called tachyons. There is no evidence that any such particles exist, and there has been much discussion of whether such a theory is really compatible with relativity. In general, physicists believe that any theory that is compatible with the theory of relativity—and hence with our experimental knowledge—must be local.

While it was true that Bohm's theory reproduced the results of quantum theory, there was a price to pay (most physicists would even say an intolerable price): it was nonlocal. In one of our conversations Bell once referred to it, more with remorse than anything else, I thought, as being "hideously nonlocal." The theory involves the instantaneous transmission of what Bohm called "quantum-mechanical forces"—new forces characteristic of his theory. In fairness to Bohm, he never claimed that his theory described rapidly moving particles, for which the theory of relativity would be essential. His theory was quite explicitly nonrelativistic, giving it, at best, a limited domain of validity. Its nonlocality casts doubts on whether it could ever be extended to include relativity. Perhaps that was another reason Einstein did not like it.

Bell put the position to me. "While Bohm had disposed of von Neumann," he said, "his theory was nonlocal. Terrible things happened in the Bohm theory. For example, the trajectories that were assigned to the elementary particles were instantaneously changed when anyone moved a magnet anywhere in the universe. I decided to find out if this was a defect of his particular picture or is somehow intrinsic to the whole situation [the situation of finding a hidden-variable theory that could reproduce the results of quantum mechanics]. I knew, of course, that the Einstein-Podolsky-Rosen setup was the critical one, because it led to distant correlations. They ended their paper by stating that if you somehow completed the quantum-mechanical description, the nonlocality would only be apparent. The underlying theory would be local [unlike Bohm's]. So I explicitly set out to see if in some simple Einstein-Podolsky-Rosen situation I could devise a little model that would complete the quantum-mechanical picture and would leave everything local. I started playing around with the very simple system of two spin-1/2 particles [the example in Bohm's textbook], not trying to be very serious, but just to get some simple relations between input and output that might give a local account of the quantum correlations. Everything I tried didn't work. I began to feel that it very likely couldn't be done. Then I constructed an impossibility proof."

If one were not worried about locality, one could imagine a mechanism that would reproduce the quantum-mechanical correlations without "spooky actions at a distance." One could introduce a new force that would be sensitive to the orientations of both magnets. Hence after the electrons were "born" in the radioactive decay, this force would keep track of the orientations of the respective magnets and adjust the spins of the electrons in just such a way as to reproduce the observed correlations. This arrangement is nonlocal, because in order to keep up with the changes in orientations of the magnets, before the particles arrive at them, it must keep ahead of the particles. If one magnet is on the Moon, say, and the other close at hand, then the force must get to the Moon faster then the electron that is heading in that direction, no matter how close that particle is to the Moon. Thus the force responsible for this mechanism must be transmitted instantaneously.

Bell's impossibility proof comes down to showing that this is the *only* kind of mechanism that can account for the correlations. No *local* mechanism of this type will work. We cannot have Einstein's local realism *and* the quantum theory. We can have only Einstein's local realism *or* the quantum theory. As Bell put it in his paper, "In a theory in which parameters [such as might correspond to new forces] are added to quantum mechanics to determine the results of individual measurements, without changing the statistical predictions, there must be a mechanism whereby the setting of one measuring device can influence the reading of another instrument, however remote. Moreover, the signal involved must propagate instantaneously, so that the theory could not be Lorentz invariant [this is another way of saying that it could not be consistent with the theory of relativity]."

Furthermore, Bells' theorem is quantitive in the sense that the distinction between Einstein's local realism and the quantum theory becomes a testable proposition. The local realistic theories lead to correlations that are different from those of the quantum theory. This difference is known as "Bell's inequality," and it is something experimenters can test. As Bell remarked in his paper, "It requires little imagination to envisage the measurements involved actually being made." It is usually much easier for a theorist to "envisage" a measurement than it is for his experimental colleagues to carry it out. In the case of Bell's inequality, it was five years before the matter was taken up by the experimenters.

Part of the reason for this delay certainly had to do with the fact that at the time, studying the foundations of quantum mechanics was not a "fashionable" thing to do. While Bell was a very highly regarded elementary-particle theorist, this side of his work was regarded with some indulgence—Bell's *violon d'Ingres*—something that, as far as I could tell at the time, he accepted cheerfully. Perhaps he felt, instinctively, that the tide would turn.

To compound matters, Bell published his paper in a relatively obscure journal, *Physics*, which did not survive much after 1964, the year his paper appeared in it. In 1964 the *Physical Review*, published by the American Physical Society, was the leading physics journal in the world. It would have been the natural place to publish any paper to ensure maximum exposure. The problem was that the *Physical Review* had, and has, page charges. At the time, even an article of only a few pages could cost the author (or, much more likely, his or her institution) a hundred dollars or so; now it is substantially more. Bell felt himself to be a guest at Stanford, and as he told me, "I was embarrassed to ask them to pay for my article." On the other hand, *Physics* actually *paid* authors to publish in it. "I thought," Bell said "that I would submit my paper to *Physics* and that that would be a good way to avoid embarrassment. I didn't make much money on it. It turned out that, although they paid for the article, they didn't give you any free reprints. What they paid me was just enough to buy some reprints."

How much more quickly Bell's paper would have been noticed if it had been published in the *Physical Review* is impossible to say, but it certainly could not have received less notice. Bell does not remember anyone's paying any attention to it at all until 1969. That year he received a letter from a physicist name John Clauser, who was at the University of California at Berkeley.

Clauser and some colleagues, it turned out, had invented a generalization of Bell's inequality which could be tested using correlated pairs of light quanta rather than electrons—a good thing, since precision experiments using properties of light quanta are much easier to carry out than similar experiments using electrons or other spin-1/2 particles. The property of light quanta that these experiments take advantage of is what is called their polarization. This is the photon's equivalent of the electron's spin. In classical terms, when a light wave propagates, the wave oscillates at right angles to the direction of propagation. Imagine making a wave motion in a rope by agitating the rope up and down. The wave propagates *along* the rope, but the wave motions of the rope are up and down, perpendicular to the direction of propagation of the rope. In the experiment that Clauser was suggesting, the spin-zero molecule radiates two light quanta, rather than two spin-1/2 atoms. The polarizations of these light quanta, it turns out, are correlated according to quantum mechanics. This is just Bohm's version of the Einstein-Podolsky-Rosen experiment again, in which light quanta and their polarizations have replaced atoms and their spins.

Clauser, along with Michael Horne, Abner Shimony of Boston University, and Richard Holt of the University of Western Ontario, published these ideas in 1969 in a paper entitled "Proposed Experiment to Test Local Hidden-Variable Theories." This paper was published in the *Physical Review Letters*, the rapid publication journal of the American Physical Society. This journal is probably the most tightly refereed physics journal in the world. It is reserved for the rapid publication of very significant new work. The fact that this paper was accepted for the *Letters* was an indication of a change in attitude toward reopening what had previously been assumed to be settled questions on the foundations of quantum mechanics. In 1972 Stuart Freedman and Clauser published, again in the *Letters*, the result of the first precision experiment that tested Bell's inequality. The result agreed with quantum mechanics and seemed to rule out all local hidden-variable theories.

This work began an era of the precision testing of the foundations of the quantum theory that continues to the present day. One of the most frequently cited examples of this work was a series of experiments carried out in Paris by a group led by Alain Aspect. A novel feature of the Aspect experiments is that the angle between the polarization detectors is switched every hundred millionths of a second while the photons are in flight. This rules out the possibility of communication between the detectors while the polarized light quanta are moving toward them. These experiments and the ones that have succeeded them also agree with quantum mechanics. The evidence is now overwhelming that Einstein's program to "complete" the quantum theory with a local deterministic theory was misguided. Local realism simply does not work.

Confronted with these results, many physicists react the way Pauli did to the earlier discussions of Einstein's qualms about the quantum theory. In 1954, in a letter to Born, he commented, "As O. [Otto] Stern said recently, one should no more rack one's brain about the problem of whether something one cannot know anything about exists all the same, than about the ancient question of how many angels are able to sit on the point of a needle. But it seems to me that Einstein's questions are ultimately always of this kind." In the present context, the results of these experiments have persuaded many physicists that the matter is settled and no longer worth discussing.

However, a growing number of physicists, confronted by just how odd quantum mechanics is, have begun to wonder whether something is missing after all—perhaps not local hidden variables, but something. No one has ever expressed this feeling better than Richard Feynman. In 1982 he said in an interview, "We always have had a great deal of difficulty in understanding the world view that quantum mechanics represents. At least I do, because I'm an old enough man that I haven't got to the point that this stuff is obvious to me. Okay, I still get nervous with it . . . you know how it always is, every new idea, it takes a generation or two until it becomes obvious that there's no real problem. It has not yet become obvious to me that there's no real problem, therefore I suspect there's no real problem, but I'm not sure there's no real problem."

One of the more puzzling things about all of this, at least to physicists, and not least to Bell, is how rapidly these very abstruse ideas, however garbled, have entered into general popular culture. An astonishing number of people who have no apparent interest in science in general seem to have heard of the Einstein-Podolsky-Rosen experiment and Bell's theorem. It is difficult to figure out how exactly this happened, but certainly books such as *The Dancing Wu Li Masters* by the journalist Gary Zukav, published in 1979 (after the first round of experiments involving Bell's theorem had been carried out), played an important role.

The last part of Zukav's book centers around Bell's theorem. At one point Zukav even notes that "some physicists are convinced that it is the most important single work, perhaps in the history of physics." When I tried this sentence out on Bell he replied, his lilting Irish accent sounding even more musical than usual, that Zukav's book was "too breathless. It gives the wrong impression of what is happening in physics institutes. People are not all desperately discussing Buddhism and Bell's theorem and the like. A precious few are doing that, whereas his book gives the impression that that's what we're doing all the time. On the other hand, it is a book in which some of the *biggest* questions that physicists discuss were brought before the public. I don't resent its success."

What seems to have excited Zukav, and others, is a train of reasoning that runs as follows: Einstein-Podolsky-Rosen correlations, acting over great distances, require an "explanation." Bell's theorem and the resulting experiments show that no explanation is possible with local physics. Therefore, relativity or not, there must exist nonlocal physics. In short, there must be "superluminal" phenomena—i.e., phenomena that propagate at speeds greater than that of light.

The lure of superluminal phenomena on the minds of the susceptible is difficult to exaggerate. It has become the late twentieth-century's equivalent of the elusive perpetual motion machine. Something of the genre is exemplified by a letter quoted in an article by Cornell physicist David Mermin, who has written a number of particularly lucid semipopular articles explaining Bell's theorem to both the general public and his colleagues. Mermin reports that the following communication was sent from the executive director of a California think tank to the Under Secretary of Defense for Research and Engineering. It reads,

> If in fact we can control the faster-than-light nonlocal effect, it would be possible . . . to make an untappable and unjammable command-control-communication system at very high bit rates for use in the submarine fleet. The important point is that since there is no ordinary electromagnetic signal linking the encoder with the decoder in such a hypothetical system, there is nothing for the enemy to tap or jam. The enemy would have to have actual possession of the "black box" decoder to intercept the message, whose reliability would not depend on separation from the encoder nor on ocean or weather conditions . . .

It is not without irony that the only actual military work that Einstein did during either world war was a brief stint during World War II working for the United States Navy on submarine detection.

Bell feels a sense of responsibility about the impact his work makes on nonscientists. He is very concerned that such people do not get the impression that quantum mechanics has implications in areas such as holistic medicine or telepathy, where there is no evidence that it does. For this reason he takes time, something that not every physicist would do, to answer questions from laypersons about such matters. While I was at CERN, he showed me a published exchange of letters that he had with University of Pittsburgh biologist R. A. McConnell, a proponent of parapsychology. In response to an inquiry, Bell wrote, "As I understand your letter, you would like me to express an opinion on the relevance or irrelevance of quantum nonlocality for parapsychology. Of course it is easy to see why people have felt there might be some connection. What has been brought out in the theoretical and experimental study of quantum nonlocality is that nature is much more curiously connected up than could have been envisaged by classical physicists, and even more so than was realized by many quantum physicists. But it must be stressed that what has been brought out here is just a feature of orthodox quantum mechanics—which fully predicts the classically inexplicable correlations which experimenters have found. Now while orthodox quantum theory is less well formulated theoretically than I would like, it is rather unambiguous in practice. And it implies rather clearly that these queer correlations do not enable us to act at a distance, for example signal faster than light. So it seems to me that, strange as quantum mechanics is, and strange as psychokinesis is, the first does not help to explain the second. Such phenomena would require, I think, the revision of quantum mechanics as well as classical physical theories."

Bell ends his letter to Professor McConnell on a conciliatory note. "But I remember how as a student in Ireland I was required to attempt experiments in electrostatics—and formed the opinion that electrostatics could never have been convincingly discovered in my home country—because of the damp. It may be that parapsychological phenomena are erratic only because of some factor analogous to damp which remains to be identified and controlled. However that may be, I would not think it sensible to ignore evidence simply on the ground that the phenomena in question would not be explicable, as far as I could see, in the context of contemporary physics and physiology. I am inclined to think that even physics is in its infancy, and that entirely new things will be found, if there is time."

Bell has also occupied himself, on occasion, with exploring the connection (if any) between the quantum theory and Eastern religions, something discussed, for example, in Zukav's book. Many of the founders of the quantum theory, such as Bohr, Pauli, and Schrödinger, had deep interests in Eastern religion. Schrödinger, in fact, wrote a brief book entitled *My View of the World*, in which he described his rather mystical Eastern beliefs. However, he took great pains in the preface to point out that those beliefs had nothing to do with the quantum theory. He said, "I do not think that these things [quantum mechanics and the rest of modern physics] have as much connection as it is currently supposed with a philosophical view of the world." Bell has had the chance to discuss these connections, whatever they are, with two of the most prominent contemporary representatives of Eastern religions, the Dalai Lama and the Maharishi Mahesh Yogi, although in the case of the latter "discuss" may not be quite the appropriate word.

On August 30, 1983, the Dalai Lama visited CERN. As Bell explained the visit, "CERN is now one of the monuments of Europe, so from time to time, important people come to pay their respects, and if they're important enough, they are received in state. An the Dalai Lama was one of those." In his book *Seven Years in Tibet*, Heinrich Harrer, who was the Dalai Lama's tutor when the Dalai Lama was fourteen, observed that as a boy the Dalai Lama was fascinated with the workings of technological devices, such as movie projectors. He learned, for example, with no instruction from Harrer, to take apart and reassemble the one Harrer found for him in Lhasa. After his visit, the *Cern Courrier*, the laboratory's house journal, ran a delightful photograph of the Dalai Lama seated at the controls of one of the huge CERN particle detectors with an impish smile on his face.

I asked Bell whether, in that vein, he thought that the Dalai Lama might have been aware of some of the connections people had been making recently between quantum physics and Buddhism. "I think that might have been an element in it," Bell answered. "When he came there was in his party an Englishman named David Skitt. He clearly knew about these things and was interested in them. He is some kind of a curator of a Tibetan-Buddhist monastery near Vevey, on Lake Geneva. That brings him to this area several times a year. I wonder, now that I think of it, if he was the one who originated the idea of the visit. In any event, the Dalai Lama came to see what was here and to learn a bit about the work. And then there was this lunch. All the important people of CERN were there to meet him. Because of this idea that maybe quantum mechanics and Buddhism have some connection, I was asked to join the party."

Bell went on to describe this extraordinary encounter. "There were thirteen or fourteen Buddhists [robed] on one side of the table and thirteen or fourteen CERN people [in business suits] on the other side of the table, and we chatted. A somewhat formal chat, in that the Dalai Lama did not admit that he spoke English on this occasion. I suspect that was in order that he would have time to think about everything. Such a man must be careful of what he says. I have heard him on television speaking English fluently, but on this occasion he insisted on doing everything through an interpreter, both listening and speaking. He raised some very interesting questions in that they came from such a different angle—so they are odd questions. For example, he said that he could not believe in point elementary particles, because he did not see how if you put together a lot of points you could ever make an extended object. It didn't seem to have occurred to him that there could be spaces in between the points. Curious things like that."

Bell shook his head and went on, "He was also very interested in the Big Bang theory, according to which the world had a start and probably will have an end. This appears to be somewhat contrary to Buddhist scripture, which emphasizes eternal recurrence; things happen again and again. I pressed him on that. Of course, I insisted that the Big Bang was a fashion in science that could change. But if science did become committed to a one-time universe, how could that be reconciled with Buddhist scriptures? He listened through his interpreter and replied, 'Well, it is perhaps not part of the Buddhism to which we are completely committed. We would have to study our scriptures very carefully, and, usually, there is some room for maneuver.' 'Some room for maneuver' was the phrase the translator used. I liked that very much.

"Another thing I brought into the conversation was the question of whether Buddhism and science really go together historically. If you read the books of Joseph Needham on science and civilization in China, you find that Needham tends to insist that Buddhism is hostile to science, and that science had a hard time taking hold in countries that were dominated by Buddhism. I have the impression that that is true. Buddhism is an inward-looking religion, concerned with personal salvation. Once you decide that that is the really important question, then questions of how stones accelerate as they fall, and so on, are ones to which you are not going to give much attention. I brought this up. To my surprise, most of the Buddhists had apparently not heard of Needham. It was a difference of perspective. For me, he is the great bridge between Eastern and Western science. And they did not know about this great work of scholarship. But the Englishman, David Skitt, knew about it. He maintained that Needham was consistently wrong in his view of Buddhism. Skitt and I got together afterwards, and we have become great friends. I see him whenever he comes to Switzerland. We carry on this discussion, but we don't force it to a conclusion. I don't know whether he would still insist that there is nothing in what Needham says, because it seems clear to me that there is."

I asked Bell if he and the Tibetans discussed the putative connections between Buddhism and the quantum theory. "We didn't discuss that too much," Bell answered regretfully. "I had the impression that they were not all that interested. They were very hazy about any kind of physics and especially about modern quantum mechanics. I think, for them, these analogies are just not very important. Physics is a subject which changes rapidly and is concerned with a very limited set of phenomena in the world. I think it would be a bit absurd to try to find the support for some big philosophy, or religion, in anything so ephemeral and specialized as that. I doubt that they have analyzed it to that degree. I think that they have just not paid attention to physics very much. They don't feel they need any support for their system. They are not looking around, grasping for straws. They are quite self-confident in their tradition. I don't think they were particularly interested in seeing me. On the other hand, I was particularly interested in seeing them, because I wanted to know what Buddhists think about these things. And especially people who have grown up in Buddhism, rather than people from the West who are going around, looking for a solution, and have decided to try this one."

"It must have been an amazing occasion," I said to Bell, wishing I had been there. "It was," he answered. "The only parallel in my experience was the time I met the Maharishi. He has an international university, and they sponsored a little symposium in a place called Vegas, near Lake Lucerne. The symposium was on physics, and the implications of physics for religion, and so forth. I was invited, among other people.

"The Maharishi was a much more regal figure than the Dalai Lama. He sat on a sort of white throne, in his white robes, surrounded by about thirty acolytes, mostly ladies, also in white robes—forming a kind of audience—looking very sweet, but saying nothing. In the middle of the gathering—which for a scientist is quite an uncomfortable atmosphere of adulation—were the handful of people who had been invited to discuss these problems. We all made little speeches, and so I made a little speech. My speech was, of course, very skeptical in character. He had, at that time, as head of his physics department, a man called Larry Domash. Domash was trying to see some analogy between the state of lowest energy of a superconductor and the state that people reach in meditation. I expressed great skepticism about that also. Domash would occasionally ask the Maharishi for his view—invited pronouncements. There was no discussion. For me he was just a figure on the throne making pronouncements.

"I was shocked to learn in the course of that, that he thought he could make rain. 'The sky is blue,' he said. 'There is no cloud anywhere. Then you relax, and a little cloud appears. It grows and grows, and soon there is rain.'

"I liked the Maharishi setup," Bell concluded, "because it was vegetarian. The meals were very good."

While many physicists take comfort from the fact that the experiments have vindicated quantum mechanics and ruled out Einstein's local reality, Bell does not. For him, it only deepens the mystery. As he explained it to me, "The discomfort that I feel is associated with the fact that the observed perfect quantum correlations seem to demand something like the 'genetic' hypothesis [identical twins, carrying with them identical genes]. For me, it is so reasonable to assume that the photons in those experiments carry with them programs, which have been correlated in advance, telling them how to behave. This is so rational that I think that when Einstein saw that, and the others refused to see it, *he* was the rational man. The other people, although history has justified them, were burying their heads in the sand. I feel that Einstein's intellectual superiority over Bohr, in this instance, was enormous; a vast gulf between the man who saw clearly what was needed, and the obscurantist. So for me, it is a pity that Einstein's idea doesn't work. The reasonable thing just doesn't work."

For Bell, the problem remains just what it was when he first began learning about the quantum theory in Belfast some forty years ago. Where does the quantum world stop and our world begin? "What worried me then," Bell reminisced, "was how to get rid of that division. I was looking for some reformulation of the theory that would permit its elimination. It was clear to me then that the hidden-variable approach would be one such formulation. If you gave definite properties—'hidden variables'—to the elementary quantum particles, you don't have to be concerned that the classical apparatus has definite properties. *Everything* has definite properties. It is just that they are more under our control for big things than for little things. It was clear to me that that was one line of possible development. I am not really looking for revisions in the standard quantum-mechanical calculations. I think that when we have solved the problem of the interface between the classical and the quantum-mechanical worlds, there will be something different in the theory. On the other hand, I am not like many people I meet at conferences on the foundations of quantum mechanics. There are many people there who have not really studied the orthodox theory. They have devoted their lives to criticizing it, and to thinking of revisions of it. I think that that means that they haven't really appreciated the strengths of the ordinary theory. I have a very healthy respect for it. I am enormously impressed by it. So I'm not thinking it will be swept away. But I am thinking that, nevertheless, some aspects of it may be changed."

Bell got up from his chair and walked around a bit. Then he said, "Doing quantum mechanics may be something like riding a bicycle. I am not sure that there is anybody who knows how to ride a bicycle. From time to time I come upon quite complicated papers on the stability of a bicycle. When you read them, you find that the situations they treat are highly simplified—like having no rider on the bicycle. Nonetheless, bicycles are rideable. I suspect that the complicated interactions that you have in bicycle riding between man, machine, and mind are maybe a bit like what we do when we do quantum mechanics. We don't quite know what we are doing, yet we can do it. We get on, and after a time we find that we can stay up, and write papers, and so on. Part of the reason is that we tackle suitable situations where the division between the quantum and classical worlds is clear. We select our problems so that the division is plain. No one would be so crazy as to regard one atom, in a hydrogen molecule, as a classical apparatus, observing the atom. The division between the quantum system and the classical apparatus has to be placed with good taste."

When he finished his pacing, Bell sat down and said, "Perhaps I did something to rekindle interest in these questions. People who are younger than me now tend to agree that there are problems to be solved. Of course, most of them don't tackle these problems. They rather work on lines in elementary-particle physics like string theory. But they are generally more open to the idea that there are problems with the foundations of the quantum theory than their teachers were."

On a bright, sunny early winter morning, Bell and I decided to take some time off from discussing quantum mechanics and go visit the huge electron-positron collider which is known simply by the acronym LEP, for "Large Electron Project." Mary Bell wanted to come along as well. She had been involved in designing one of the components of the machine, the so-called "electron-positron accumulator"—a little ring in which the electrons and positrons are accumulated and stored before they are swept by electromagnetic fields into the big machine. She had not yet seen the whole machine, which was nearing completion. So Mary drove out from Geneva to meet us.

As it happens, another old CERN friend of mine, an Italian physicist named Emilio Picasso, is the director of the LEP. When I called to ask if we could have his permission to visit the machine, he said he would be delighted, but was too busy to show us around himself. He would arrange, however, for the senior administrator of the Project, Manfred Buhler-Broglin, to be our guide. We all agreed that we would meet at ten in the morning in front of the CERN cafeteria.

A little before ten, Bell and I went down to the cafeteria for an espresso, and a little later Mary joined us. Both Bells were wearing whitish ski parkas and slacks, and Bell had on what looked like an Icelandic fisherman's sweater, with brown snowflake designs. I had not seen Mary since her retirement. After we greeted each other, I remarked that I had heard from Bell that she will not look at any more mathematics. "He is exaggerating," Mary said with a hearty laugh. Bell looked relieved.

After a few minutes, Picasso and Buhler-Broglin appeared. Picasso introduced me to Buhler-Broglin, who has a serene round face and was the only man wearing a tie. One of his most important jobs, I later found out, was community relations. The construction, which has gone on for several years, has been quite disruptive to the lives of the small French communities near which the twenty-seven-kilometer LEP tunnel runs. Buhler-Broglin has attended all sorts of community meetings to try to ease the relations between the laboratory and these communities, and to make sure that the completed machine leaves as little impact as possible on the environment. The ring through which the electrons and positrons run is buried deep underground—in places, some five hundred feet—so there is no significant problem with radioactivity.

Our intention was to visit a couple of the sites where access to the buried tunnel is possible. Winding through the entire twenty-seven-kilometer tunnel is the metal evacuated pipe, within which the electrons and positrons will run in opposite directions. The pipe has a complicated, roughly ellipical shape, with dimensions of something like six by two inches. The beams of electrons and positrons within it have dimensions roughly the size of one's thumb.

Shafts have been sunk into the ground at a few locations so that people (technicians, experimenters, visitors like us) can go down into the tunnel. As it happens, these locations are in France, several miles from the main CERN laboratory, so Buhler-Broglin went to get a CERN staff car to take us over the border. When he came back, we bundled ourselves into the small car and headed for the customs checkpoint, which is just outside the entrance to CERN. I had forgotten to get a French visa, which was then necessary, but fortunately the lab was able to provide special temporary visas for visitors to the LEP. The electrons and positrons, which will cross the border several times on each circuit around the tunnel, will have no such problem.

After driving for some time in the delightful French countryside, we left the main road and stopped at something that looked, at first sight, as if it might have been the surface operation of a mining company. This impression was reinforced when Buhler-Broglin produced three yellow plastic mining helmets for us to wear. Mary got her abundant curls under her helmet, while Bell's red beard protruded from his, giving him a slightly nautical look. Buhler-Broglin had an official-looking blue helmet.

We were then taken to an elevator. It descended five hundred feet down into the earth. My only comparable experience was when, a couple of years ago, I visited what is known as the IMB (for Irvine, Michigan, and Brookhaven) detector at the Fairport mine in Painesville, Ohio. It is located *two thousand* feet down a working salt mine owned by the Morton Salt Company. It was this detector, a tank of ultrapurified water, that detected the neutrinos that came from the supernova explosion of February 1987.

In the case of the LEP, the detectors are huge electronic scanning devices, the size of a three-story house, that can be wheeled in and out of the beam on railway tracks. Using detectors like this, and a lower-energy proton-antiproton collider, the experimenters at CERN in 1982–83 made one of the most important discoveries in contemporary elementary-particle physics: they found the carriers of the weak force that is responsible for radioactivity. Unlike the light quantum, which carries the electromagnetic force and is massless, these quanta are very heavy on the scale of elementary particles, the electrically neutral specimens weighing something like ninety times the mass of the proton—as much as a sizable atomic nucleus. To find these carriers, the CERN experimental terms, which had typically a hundred and fifty or so people in them, had to scan a *billion* collisions in a thirty-day period in order to find five—*five*!—useable events. When the LEP is fully implemented with its complement of superconducting magnets, it is expected to produce a *thousand* of these events an *hour*. This will open up a whole new era of high-energy experimental physics.

We reached the bottom of the elevator shaft. Several groups of neatly dressed scientists and engineers were going about their jobs, speaking the many languages of the people who make up the CERN community—French, German, Spanish, Italian, the Scandinavian languages, Russian, English, Chinese, and Japanese. The whole thing, even the languages, gave me a great sense of order. I could see the LEP tunnel. It is so long that when one looks down it one does not have any sense that it is curving. So much human effort spent to uncover—to reveal to us—this beautiful, austere quantum world. Somewhere that world stops, and ours begins. But where?

My visit to the LEP with John and Mary Bell took place in January 1989. The machine was officially scheduled to be turned on on Bastille Day the following summer. Bastille Day had been chosen as a gesture to the French, who had donated much of the land. As it turned out, because of objections by Mrs. Thatcher (Britain had contributed a share of the money), the official opening took place a little later.

The machine worked perfectly and continues to do so. The experimental groups, whose building-sized detectors we had visited, began taking data soon afterward and by the late summer announced results on the mass and "width" of the Z^0.

The fact that the Z^0 has a width is something one can trace back to the Heisenberg uncertainty principle that connects energy and time. The Z^0 is an unstable particle with a lifetime of less than 10^{-24} seconds. Since the accuracy of an experiment in which an energy is measured is limited by the Heisenberg relation $\Delta E\,\Delta\tau \approx \hbar$ (where $\hbar$ is Planck's constant), this means that an unstable particle can never have a precisely determined rest energy or mass. In the present instance, each production of a Z^0 in an electron-positron collision produces a Z^0 with a slightly different mass. The spread of these masses is what is known as the width of the Z^0. This width can be calculated from quantum mechanics if one knows the decay mechanisms. The Z^0 has several possible decay modes. About ten percent of the time it decays into particles such as electron-positron pairs of neutrinos, and the rest of the time it decays into quarks.

The experimenters choose a convenient decay mode in order to measure the width. Quantum mechanics tells us that any mode will manifest the same width as any other. The particle has a single, unique lifetime.

By comparing the observed width with the theory one can learn about decay modes that cannot be directly measured. For example, it is not possible in these experiments to directly measure the decay of a Z^0 into, say, a pair of neutrinos (these interact so weakly that they simply escape from the detector). But one can tell that the Z^0 *does* decay into neutrinos, because if one leaves out neutrino decays in the theoretical calculations the predicted width is too small. It disagrees with the experimental result. Moreover, one can also learn how many different *types* of neutrinos the Z^0 decays into, because the number of types determines the width.

The CERN groups found that the Z^0 can decay into only three types or "families" of neutrinos. This was both a very mysterious and a very satisfying result. It was satisfying because the cosmologists have been telling us for some time that the amount of helium produced in the early universe is also determined by the number of neutrino families. The cosmodynamics of the early universe is related to the energy density in the early universe, and this in turn is related to the number of neutrinos that were present. These same cosmologists have been insisting that having more than three families would have produced an uncomfortably large amount of early-universe helium. Thus the fact that the terrestrial accelerator experiments also limit the number of families to three is very satisfying. But it is also very mysterious. Why precisely *three* families—and not four, or seven? No one has a good explanation. Perhaps the explanation will take us into the interweavings of quantum mechanics and cosmology—the domain of the subject of the next of these profiles, John Wheeler.

John Wheeler

RETARDED LEARNER

IN THE FALL of 1983 I audited one of the most remarkable and idiosyncratic physics courses I have ever encountered. It was called Foundation Problems of Physics and was open to graduate students only, so a solid background in advanced physics was assumed. It was taught by John Archibald Wheeler, then (like me) an I. I. Rabi Visiting Professor at Columbia University. (At the time, Rabi was still alive, and Wheeler and I were able to tell him how much we enjoyed occupying his "chair.")

Wheeler was then seventy, but extremely active. He was the full-time director of the Center for Theoretical Physics at the University of Texas at Austin. That semester he commuted back and forth between New York City and Austin, with numerous stops elsewhere to fulfill various lecture commitments—Wheeler is a very popular lecturer. When I asked him about his almost incredible energy, he attributed it partly to the fact that he managed to swim a quarter of a mile a day in a small enclosed pool by his house in Austin.

A few years after he gave the course, I had several lengthy conversations with Wheeler about his life and work, and I took the opportunity to ask him how the course had come about. He told me that the title had been invented by Gerald Feinberg, then chairman of the Columbia physics department, who had invited Wheeler to visit the university. The only trouble was that Feinberg neglected to explain what the course was to be about. So, Wheeler explained, when it began he had no clear idea of how to go about teaching it. However, halfway through the first hour, Wheeler recalled, "I got this idea." He asked each of the twenty-odd students to turn in, after each lecture, a single summary sentence that they felt best summarized the lecture, along with what

Wheeler characterized as a single "pregnant" question. Wheeler would study these papers carefully, and in the next class would hand out neatly handwritten mimeographed notes with the class' contributions and his own responses. The class became a little like the "surprise" version of the game 20 Questions, in which the participants have no preset object in mind, but rather the object is created by the questions as the game proceeds—an image, incidentally, Wheeler sometimes uses in characterizing the quantum theory of measurement.

Many of the questions had to do with relativity and cosmology, subjects closely identified with Wheeler's work. Typical were questions like "Why did the Big Bang occur?"—to which Wheeler responded, "Physicists' standard way to avoid misunderstanding of the word 'why': translate to, *How* did it come about that there was a Big Bang?" To which Wheeler added, "A *wonderful* question, worthy of thought every day."

Some of the questions had a philosophical bent, which did not seem to faze Wheeler. For example, one student asked, "Where is the barrier between solipsism and objective reality in a physical theory?" Wheeler responded to this in the slightly skew way that sometimes characterizes his discourse: "The Los Angeles girl locked from babyhood to age 13 (when the neighbors found out and called the police) in an attic room, given food but never spoken to, had by that time lost the power not merely to speak, but even to think. There is not a word we utter, a concept we use, an idea we form, that does not directly or indirectly depend on the larger community for its existence. 'Meaning'—and what else is 'objective reality' if it is not meaning? . . . is the joint property of those who communicate."

A few of the questions were beyond Wheeler. One student asked, "Does the similarity between different equations of physics mean the consistency of the physicist's picture of the world (which is exciting to us) or merely the *heredity* of physical ideas (which is less exciting)?" To this Wheeler responded, "I am too stupid to know how to analyze this question."

To me, the most interesting questions concerned the interpretation of the quantum theory. These students were just confronting the theory fresh. Like most gifted students, they found the mathematics straightforward. It is not the mathematics that is the problem (something that is difficult to communicate to the layperson for whom anything involving mathematics is, almost by definition, difficult); it is what the mathematics is supposed to mean—the sort of thing that so much bothered Bell when he was a student. In the questions these students asked, one saw all the difficulties in understanding the quantum theory, and in Wheeler's answers, all the difficulties in responding to these questions. Here are a few samples quoted at some length.

The first question had to do with what is meant when one claims in quantum mechanics that one knows—can predict in advance—the "state" of, say, an atomic electron. In responding, Wheeler attempted to clarify for the student what is meant by the "state of an electron" and what is meant by its determinability. He wrote, $\psi_7(x)$, "The wave function of an electron in, say, the 7th quantum state, we can know perfectly from our calculations based on Schrödinger's equation, and in a common and generally accepted way of speaking this means that we know the 'state' of the electron." In other words, the solutions to the Schrödinger equation—the wave functions—are just as causally deterministic as the orbits of classical mechanics. But these wave functions describe only the probable outcomes of experiments to be performed. As Wheeler went on to explain, "What we do not know and ordinarily cannot know [in quantum mechanics] is 'the value' of a dynamical variable [such as position or momentum] until (1) We or, better, our apparatus *decides* which of one or another complementary variables to measure and (2) 'Nature,' in the shape of an "irreversible act of amplification," gives us an answer. Only then do we know the position of the electron or through which slit the photon came, etc. But the use of [the term] 'state' to describe *that* information, while understandable among friends who make allowances for slurring of terminology, is truly dangerous in the larger world, where people have such a tendency, an understandable tendency, to *misunderstand*!"

In this answer Wheeler made use of a phrase taken from his teacher Niels Bohr, a phrase that he was to repeat often during the course—namely, Bohr's dictum that "no quantum phenomenon *is* a phenomenon until it has been registered by an irreversible act of amplification." Sometimes Wheeler illustrated this with three baseball umpires discussing their craft.

> No. 1: "I calls 'em like I see 'em."
> No. 2: "I calls 'em the way they *are*."
> No. 3: "They ain't *nothin'* till I calls 'em."

In a homey way, No. 2 (Einstein) and No. 3 (Bohr) are a perfect summary of the issue in that great debate. It was interesting to me that the students did not find Bohr's formulation entirely satisfying. Many of them wanted to know what words like "amplification" and "irreversible" really meant. One student asked, "What exactly is the meaning of 'registered'? 'Irreversible act of amplification'? 'Indelible'? And why does quantum mechanics *need* such a *classical* process, whereas classical theory is only a limiting case of quantum theory?" Wheeler's answer, which was, in its entirety, "Bohr: To be able to *tell* one another what we found," would have driven the young Bell, or even the present Bell, to distraction. To a second student, who asked, "What is the definition of registered? How without it can one properly define elementary quantum phenomena?" Wheeler responded it meant "Brought to a close 'by an irreversible act of amplification' (blackening of a grain of photographic emulsion, click of Geiger counter, etc.) was Bohr's version. That it is 'open ended' is an inspiring indication that there's something great yet to be learned!"—a sentiment that would certainly be endorsed by Bell.

As this course and my subsequent conversations with Wheeler made clear, he thrives on students. As he once remarked to me, "I am one of those retarded learners . . . one of those people who can't learn except by teaching." He has an unusual lecture style. It features diagrams that appear to grow organically as the lecture proceeds. One has the feeling that they will continue to evolve on their own after everyone has left the room. They are at once enormously dense and yet very clear—like Dali drawings. Perhaps this can be attributed to Wheeler's early training in engineering drawing at Johns Hopkins University or to an even earlier course in drawing he took when he was still a high school student in Youngstown, Ohio. Wheeler thinks, and has always thought, in a very pictorial way. He remembers as a young child being given a book by his father called *Ingenious Mechanisms and Mechanical Devices*. "Each page," he told me, "was absolutely marvelous to look at—an illustration of cleverness. I can remember lying in bed thinking of those pictures. It wouldn't be words. It wouldn't be equations. It would be pictures of how one thing would fit together with another." Wheeler's ambition then was to make a clock with little actors that would emerge when the hours chimed, but he settled for making a machine that solved algebraic equations and a sewing-machine cabinet for his mother with hinges that did not show. "That was a very fancy doing—my own design," he told me. "It had a tulip-leaf pattern at each corner. I turned most of it out on a lathe that I made out of an electric motor with a rod at the other end."

In the fifty years that Wheeler has been teaching physics—there was a substantial hiatus during the war—he has produced some fifty Ph.D.s (the most famous of whom was no doubt Richard Feynman). This is an enormous number of doctorals for a *theoretical* physicist. An experimental physicist, by contrast, can often have several Ph.D. students working together on a single experiment. For a theorist, each Ph.D. represents at least one publishable idea, and fifty is a lot of publishable ideas. In his memoir "*Surely You're Joking, Mr. Feynman!*," Feynman, who took his degree at Princeton with Wheeler in 1942 and won the Nobel Prize in 1965, has a chapter called "Monster Minds"—an affectionate reference to Wheeler. He says that when he was a student he came to Wheeler with a calculation that Wheeler, after a cursory inspection, said was wrong. Feynman writes, "What bothered me was, I thought he [Wheeler] must have *done* the calculation. I only realized later that a man like Wheeler could immediately *see* all that stuff when you gave him a problem. I had to calculate, but he could see."

In addition to supervising students, Wheeler has done an incredible amount of original research. His papers, dozens of them, are bound in several volumes, which he keeps for handy reference on a shelf near his desk in Austin. Nearby are the bound theses of many of his students. He has worked successfully in almost every branch of physics, from nuclear fission—his paper written with Niels Bohr in 1939 was the first successful theoretical treatment of fission—to black holes. The terms "black hole," "ergo-sphere," "geon," "Planck length," and "stellarator" were all invented by Wheeler.

The Wheelers trace their heritage in this country back to seventeenth-century New England. They came with the migration of the "dissenters" from the southeast of England. Wheeler told me that in 1645 there were forty-five families named Wheeler in Concord, Massachusetts. He is descended from one of them—Sergeant Thomas Wheeler. The Archibalds, Wheeler's mother's side, migrated from Nova Scotia to Kansas just before the Civil War. John Christy Archibald, Wheeler's great-grandfather, was one of the founders of Lawrence, Kansas. Kansas was a free state, while Missouri was not. In a pre–Civil War skirmish, Quantrill's Raiders—the Confederate "border bandits" under the command of William C. Quantrill—came over the border into Kansas and burned down the houses in Lawrence. They lined up Wheeler's grandfather and some other men in front of a barn door, and were about to shoot them. Wheeler's great-grandmother threw herself in front of her son, crying, "Don't shoot him. He's just a boy!" and Wheeler's grandfather escaped. During the Civil War he was nearly done in when a cannonball hit the ground in front of him, ricocheted, and grazed him in the head. He was rescued on the battlefield by a woman who took him into her house, where he revived in a couple of days.

After the war, Wheeler's grandfather matriculated at the University of Kansas in Lawrence, where he met Wheeler's future maternal grandmother, who came from Ohio. Her family had run an underground Abolitionist "railway station." Wheeler told me that "in the family we have two silver spoons. They came from a Negro woman who had lived in Louisiana as a slave and had arrived in Ohio on the underground railway to stay. She was ill, and the family looked after her for several weeks. When she left, she wanted to give my family something, so she gave them two silver spoons she had taken from the household in Louisiana. After all those years of service she thought she was entitled to take those spoons." Wheeler's father's grandfather, Ezekiel Wheeler, was a Seventh Day Adventist minister. Wheeler's father recalled visiting them and listening to heated theological discussions after they thought he had been safely tucked into bed.

After Wheeler's grandparents married, they moved to a ranch near Trinidad, Colorado. Besides running the ranch, grandfather Archibald taught school. Wheeler remembers that when he was very young, his grandfather taught him some mathematics. In all, five children were born on the ranch; his mother was the youngest. Wheeler does not count any scientists among his ancestors. However, he told me, his mother, who often walked a ten-mile round trip along the railroad tracks to school, was very gifted in arithmetic. When standing in line in a grocery store, she could look at the figures upside-down on the sales receipt of the customer in front of her and add them in her head before the sales clerk could add them right-side-up. As a Civil War veteran, Wheeler's grandfather had some sort of priority with respect to government jobs, so, when Wheeler's mother was eighteen, the family moved to Washington.

In the meanwhile, on the other side of the family, Wheeler *père* was working his way through Brown University, partly as a sign painter and partly as a librarian. Wheeler told me, "He fell in love with libraries." As a result, he studied to be a librarian at the State University of New York at Albany. After graduation, the senior Wheeler got a job in the public library in Washington, where Wheeler's mother had also taken a job. The couple met and fell in love. "Those were the days," Wheeler explained, "when everybody would gossip about everybody, so at the end of the day they would exit the library by separate exits. They would meet in Rock Creek Park and read Keats and Shelley and Byron and Wordsworth. So I was a gleam in their eye at that time." The couple was married October 1910 and moved to Jacksonville, Florida, where Wheeler was born on July 9, 1911. A few months later, Wheeler senior took a position with the Los Angeles public library, and the family settled in California, where Wheeler's brother Joe was born three years later. (He was later killed in the Second World War in the fighting in Italy.)

Not long after Joe was born, the Wheelers moved again, this time to Youngstown, Ohio. "My father had had difficulty in the Los Angeles library," Wheeler explained. "He was full of ideas of what a library should be. I have a feeling that the librarian there was rather slow-moving and felt that the whole show was being taken out from under him. One day, without any advance notice, he told my father that his position had been discontinued. The bottom dropped out of the Wheeler family. People who had been in touch with the library—solid citizens—wanted to make a big fuss, but my father said, 'No—let's just leave it.' " Wheeler's mother moved back to Washington, temporarily, with her two sons, and his father found a fill-in job at the Great San Francisco Exposition of 1914, where he met Theodore Roosevelt, who had been one of his heroes. A couple of months later he was appointed director of the public library in Youngstown, Ohio. Wheeler spent most of his childhood in Youngstown. The family was completed by the addition of a brother, Robert, born in 1917 (a geologist until his death a few years ago), and a sister, Mary, born in 1918, who also became a librarian and now lives in Vermont.

From his account of it, Wheeler's Ohio boyhood sounds like a very happy one. He had a paper route delivering the Youngstown *Vindicator* and worked in the library Saturday mornings.

Unlike many theoretical physicists I have asked (but like Bell), Wheeler does not have any special early mathematical memories. He does recall that when he was about four years old, being bathed by his mother, he asked her, "What happens when you get to the end of things?" I asked Wheeler if he had numbers in mind. "No," he said, "space."

What early scientific memories Wheeler does have, have to do with making things—comptometers, "guns" that operated with a light socket (you put something in the socket, turned on the switch, and the object popped out), and a railway signal. These activities culminated at the age of thirteen, when Wheeler and his friend Verdet Moke founded what they called the "Wheeler-Moke Safe and Gun Company." It produced wooden combination locks with, Wheeler told me, "little wheels, whittled of wood, with a little notch on each end with a pin connecting each to the next, so you could set the combination. I do remember making a little machine which would solve algebraic equations of the form $ax + by = c$ and $dx + ey = f$. There were wooden sticks on a board, and where they crossed would give the solution."

By the time Wheeler was a senior in high school, the family had moved again, this time to Baltimore, where Wheeler's father became the director of that city's public library. Wheeler attended a high school that was known as the Baltimore City College. His abilities must have been evident, since he recalls that one of his teachers at the City College, Lydia Baldwin, "went around to see my father—to say he ought to do something about me—to push me ahead. Though my parents thought I was asleep, I overheard them discussing what they were going to do about my education." As it happened, Wheeler had already skipped a few classes when, at age ten, his father had taken a year off from library work to try (unsuccessfully) to make a go of a family farm in Vermont. Wheeler had attended a one-room schoolhouse in rural Vermont, and he had been skipped ahead.

During that year he had also managed to blow off a small piece of one of his fingers with a dynamite cap. "My father," he told me, "had gotten together with the neighbors to put up the poles for a power line. That was the only way they could get electricity to the farms in those days—put in the power lines themselves. To make the holes they used dynamite, and the dynamite was kept in the loft of the pig shed. I fed the pigs there every day, morning and night, so I knew about the dynamite. I was also reading books on explosives—I loved to read about explosives. So I thought I would take a dynamite cap off and make a little explosion with it clear off across the road. I put a match in the ground and lighted it, and missed, and relighted the match and missed. This got me closer and closer, when it went off in my fingers. It blew off the end of one finger and part of my thumb. Lucky I have any fingernail left."

Besides leaving Wheeler with a slightly peculiar-looking finger, this also left him—oddly—with a highly developed taste for watching explosions. During the war, when he used to visit Los Alamos, his friends would take him out on the mesa at night when they tested their high explosives. After the war, he attended a going-away party for the Los Alamos physicist Sterling Colgate. Wheeler recalled, "He was giving away, at the farewell dinner, a case of dynamite. To me this was a lovely thing, but I couldn't see any legal way to carry it on a plane." Fireworks have always played a large role in the Wheeler family's Fourth of July celebrations.

In any event, Wheeler managed not to blow himself up before he graduated from high school at age fifteen. By this time, he had taught himself the differential and integral calculus and was working his way through a book called *Problems of Modern Physics* by the great Dutch physicist H. A. Lorentz. It dealt with both Einstein's theory of relativity and the old quantum mechanics. "It was over my head in large measure," Wheeler recalled, "but it was fascinating to me. I can't remember having the same kind of teachers in mathematics and science that I had had in Youngstown. In the physics class in Baltimore, I used to sit there working on something else. This annoyed the teacher, and he used to try to catch me out. But the class went so slowly that working on something else was the only thing to do."

At sixteen Wheeler entered Johns Hopkins University with the idea of becoming an engineer. "My three uncles," Wheeler explained, "were mining engineers. With my interests, how else was I going to earn a living except at engineering? The idea that you could do what you wanted to do, and get paid for it, never occurred to me." Wheeler's going to Johns Hopkins was, in Pasteur's notable phrase, an example of "chance favoring the prepared mind"—the "chance" being, in this case, the presence of a first-rate scientific institution in his "backyard," and the prepared mind being Wheeler's. As Wheeler put it, "We were a hard-up family, and the only place I could have afforded to go to college was in Baltimore." Wheeler's engineering career lasted one year, during which he took things like surveying, mechanical drawing, and strength of materials. He then spent the summer rewinding electrical motors in the Pittsburgh Verde Grande silver mine in Zacatecas, Mexico, of which his uncle was the manager. He found this a most unpleasant experience. Also, by this time he had discovered the physics library, which shared the same facilities with the engineering library at Johns Hopkins. While he was doing his homework on the bending of metal beams and the like, he would take a glance or two at the latest issues of the *Zeitschrift für Physik*. This was 1927, and most of the articles on the then new quantum theory were being published in the *Zeitschrift*.

Johns Hopkins had a six-year program that led directly to a Ph.D. This meant that there was no clear-cut distinction between the undergraduates in the program and the graduate students. (Indeed, Wheeler has no undergraduate degree. However, he does have a Ph.D. from Johns Hopkins, along with twelve honorary doctorates.) There were some thirty students in the physics program and classes were taught mostly in seminar fashion. Students were asked to work for a certain number of weeks with each professor so that they could get a variety of hands-on laboratory experiences. It was an ideal arrangement for someone with Wheeler's vivid intellectual curiosity. He recalls, sometime after 1929, encountering on campus the physicist Joseph Sweetman Ames, who had become the president of the university. "I don't think he knew me from Adam," Wheeler noted, "but he asked me how I was getting along. He said that the whole thing around there was to be very interested in *something*. I told him I was."

Among the things that Wheeler was interested in was the new quantum mechanics. "It was being used at Johns Hopkins," he told me, "to make predictions about spectra." The students had a kind of joint seminar with some of the faculty members in which they tried to educate each other about the new theory. There were few, if any, formal courses on the quantum theory in American universities at that time, and people such as Rabi and Oppenheimer went to Europe to learn it. Mainly, Wheeler taught it to himself. "At that time," he recalled, "the family would go to the farm in Vermont for a month in the summer. We'd sold the farm, but we had kept a few acres by a brook where we had a cottage . . . really a sort of camp. Anyway, I would sit out in a pasture with Hermann Weyl's book *Gruppentheorie und Quantenmechanik* [a classic text which Weyl published in 1928]. That was the way I learned my quantum mechanics. It was wonderful, because at that time I was learning German, and the German and the physics went hand in hand. I retained a great admiration for Weyl and feel so lucky to have known him later in Princeton."

Wheeler's thesis, which he did with the distinguished spectroscopist K. F. Herzfeld, had to do with the absorption and scattering of light in helium. "Later," Wheeler reminisced, "when I was in Copenhagen I was told a story about Bohr. Rutherford had been beating on him to publish his 'Bohr atom.' And Bohr said, 'I can't. Nobody will believe me until I do all the atoms and all the molecules.' Rutherford said to him, 'Bohr, you do hydrogen and you do helium and everybody will believe the rest.' But with the old quantum theory he couldn't do helium and here I was, for my thesis, about to do helium with the best approximations then available—to explain the refractive index and its dependence on wavelength. Thinking back on it, I cannot think of a happier subject. The thesis was no great thesis because I was pretty young and green. I didn't realize all the things that could have been written about the subject, that could have blown up into a really wonderful dissertation. Incidentally, in 1932, Einstein's friend Paul Ehrenfest came through on his way to Caltech for a visit. I can recall Herzfeld trying to get Ehrenfest to talk at the seminar and Ehrenfest saying it must be Herzfeld who talks. Finally, after a lot of persuasion, Herzfeld agreed to talk. He talked about his ideas for treating problems that do not admit the standard separation of variables in the wave equation. Ehrenfest got up and said, 'My dear Herzfeld, you are completely crazy.' It was wonderful to see the spirit of collaboration between the two of them."

Wheeler got his degree in 1933—the depths of the Depression. "A neighbor shot himself," Wheeler recalled, "because he did not see how he was going to support his wife and children. Well-dressed people came to the door asking for any kind of job—painting, mowing the lawn, shoveling coal—anything. Desperate years. My father had, nonetheless, gotten the city to appropriate money for a new public library. He went again and again to the state legislature to lobby for the money. Baltimore got an unbelievable amount for the money it spent on that library. It was the only library in American to have a storefront window with glass coming down to street level so that everybody could look in and see that a library was an attractive place and not a terrifying place."

Most young physicists had enormous difficulty finding work in their field. Although Wheeler, who is always unassuming about his work, downplays the quality of his thesis, it was sufficiently impressive that he won one of the very rare National Research Council Fellowships. "I somehow wasn't aware of all the difficulties there were then. I didn't know what I was going to do if I didn't get a job. Recently I wrote a little piece about my friends who were looking for jobs at that time. There was, for example, Larry Hafstead, who was a graduate student along with me. He couldn't get any decent job so he took a 'peanut' job at the Carnegie Institute for Terrestrial Magnetism in Washington. He went on to do wonderful work in nuclear physics and became director of research at General Motors, with a research budget of nearly a billion dollars a year. And there was John Mauchley, who couldn't get a teaching job at any important place. He took a teaching job at a one-horse college on the outskirts of Philadelphia. He went on to develop the electronic computer, which eventually became Univac and Remington Rand. There was a chap who was interested in the infrared. But nobody was interested in the infrared then; no government laboratory, no industrial laboratory, no university. Then he realized that, good times or bad, people got sick. He took his equipment to the Johns Hopkins hospital, where it was used to measure body temperatures for different parts of the body. He ended up as the director of a big laboratory at Yale devoted to the study of physiological conditions by physical methods."

Wheeler had his choice of working with Oppenheimer in California or Gregory Breit at New York University. He made the somewhat surprising choice of Breit. He had seen Oppenheimer in action at meetings of the American Physical Society, and says, "Oppenheimer was much quicker than I was accustomed to. Breit was a slower thinker. I have to ponder and ponder on things, so his style fitted me. I learned a lot of physics with him that year. He was at the uptown campus of New York University in Washington Heights and I can recall a lunch one day when one of our experimental colleagues reported on some experiments on electron scattering that didn't fit the theory. The young theorists were sitting around the table saying the experiments couldn't possibly be right since they violated the law of the conservation of parity." The experiments, which were right, and were dismissed, anticipated the discovery of parity violation in the weak interactions by some twenty-five years.

While he was a student at Johns Hopkins, Wheeler found time to lead an active social life. "I was one of the founders and the first president of the Baltimore Federation of Church and Synagogue Youth," he told me. "I had also been chairman of the graduate student committee that ran the dances for the graduate students. I had taken various girls around. The day I took my

Ph.D. oral exam I met my best friend Bob Murray and one of us said—I think it was me—'let's get Janette Hagner and take her on a walk in the park'—Druid Hill Park. So we went and got her and we took a walk. I had met her through her sister. Janette had graduated from Radcliffe and was doing some work in history at Johns Hopkins. She had also had a one-year fellowship to Rome to do Italian history. During the walk, she happened to mention that she was going to be teaching school at the Rye Country Day School, outside of New York. I took advantage of that while I was at New York University. We went out three times together and we were engaged." Engaged but not married. That came a year and a half later. "There was the problem of no money. Looking back on it, it was the craziest thing. We should have gotten married anyway and gone to Denmark together."

Every scientist—Einstein being a notable exception—can find in his or her career a decisive teacher. For Bohr it was Ernest Rutherford. For Feynman it was Wheeler, and for Wheeler it was Bohr. When Wheeler applied for an extension of his National Research Council Fellowship, he wrote on his application that he wanted to work with Bohr because "he sees further than any man alive." One of the people at NYU told Wheeler that he should wait to go to Copenhagen until he was a "mature physicist" (Wheeler was only twenty-two), but with Breit's encouragement, Wheeler applied anyway and was accepted.

Wheeler went to Copenhagen via Germany. He is still haunted by the memory of changing trains in Cologne. He found that, because of the German inflation, he was the only person in the railway station who could afford the price of breakfast. "There was a storm trooper there," he told me, "wearing boots and a swastika, striding up and down, glaring at me. On the boat I had met a chap who had had farm jobs in America. He was going back to Germany because he thought that, with Hitler, there was a future there. I always wondered what happened to him."

Bohr was forty-nine when Wheeler met him. He had, Wheeler explained to me, "two speeds—not interested or completely interested. This applied to everything. E. J. Williams, who was at

Copenhagen when I was there, told me that during the Stalin period he accompanied Bohr on a visit to the Soviet Union. Bohr was allowed to visit the Kremlin Museum with Williams. It was a tremendous privilege. It was closed normally. Bohr was just not interested in anything, and Williams felt very apologetic. His guide didn't seem to get anywhere with Bohr. Then he came to one of those old carriages. You could ride in them. Bohr became fascinated by how the springs were mounted, how you kept the vibration out. He couldn't be dragged away. It was two speeds—not interested, or completely interested."

When Wheeler arrived in Copenhagen, the thing that Bohr was completely interested in was whether quantum electrodynamics was right at high energies. Wheeler told me that before he had come to Copenhagen he had listened to "an absolutely packed evening lecture by Oppenheimer in which he said that electrodynamics must surely fail when the energies get to be about 137 electron masses [a relatively low energy by present accelerator standards]. But Bohr and Williams had developed a point of view from which you could deduce [correctly] that it could not break down at these energies. Hence the new particles that were being discovered that penetrated ten centimeters of lead could not be electrons. This created the climate of opinion that was necessary for the discovery of the meson. Williams was able to take advantage of a cross section I had worked out with Breit for the production of electron-positron pairs in the field of a nucleus. I remember talking to Bohr and Williams about it. I was so shy that I got up to erase the board with the back of the eraser. Bohr then got up, took the eraser, turned it over and said, 'It would be easier if you did it this way.'

"It was a wonderful feeling to see a cross section that seemed to have no prospect of being experimentally determined, being put immediately to use. At Easter time Christian Møller had gone from Copenhagen to Rome to visit Fermi's lab, and he came back with these reports of unbelievably enormous cross sections for the absorption of slow neutrons by nuclei. They were absolutely at variance with the idea of a particle's going freely through a nucleus, although that was not, at first, clearly recognized. But Møller presented the results. Before he had gotten fifteen minutes into his talk, Bohr took away the blackboard from him and was explaining how these results could change our whole idea about nuclear structure."

Not long afterward, Bohr developed what has become known as the "liquid-drop model" of the atomic nucleus. In this model, the nucleus is treated like a tiny globule of an incompressible fluid. One does not attempt to treat the individual neutrons and protons in the nucleus, but only their collective behavior. The model has been enormously successful, especially in the study of reactions involving heavy nuclei. Almost at once Wheeler began applying it. He was also writing to "any place I had ever heard of" for a job. A physicist named Arthur Ruark had been brought in to build up the physics department at the University of North Carolina in Chapel Hill. He had heard Wheeler give a lecture, and he hired him as an assistant professor. "The job paid, I think, $2,300 a year," Wheeler told me. "Janette gave up her job, which paid $2,500 a year, and came with me to Chapel Hill. As I told you, when I look back on it, it was the craziest thing for us not to have gotten married before I went to Denmark and to have gone together." The Wheelers have been married close to sixty years.

When he went to Chapel Hill, Wheeler had every intention of remaining there forever. His two elder children, Letitia Wheeler Ufford and James English Wheeler, were born there in 1936 and 1938, respectively. (His younger daughter, Alison Wheeler Lamston, was born in Baltimore in 1942.) However, in 1938 Wheeler was offered a position at Princeton University. The distinguished theoretical physicist E. U. Condon had just left the department, and the university had decided to replace him with two nuclear theorists, Eugene Wigner and Wheeler. A program of experimental nuclear physics had also gotten under way at the university.

Neither Wheeler nor Wigner was a stranger to Princeton. Wigner had already taught there and, not getting tenure, had gone to the University of Wisconsin before being called back. Wheeler had first visited Princeton in 1933 when he went there from New York to attend the first lecture Einstein gave at the newly created Institute for Advanced Study. (It did not then have its own quarters, but shared Fine Hall with the physicists and mathematicians from the university.) Of the lecture, which had to do with Einstein's ideas on a unified field theory, Wheeler commented to me, "Einstein was not the kind of retail dealer of equations I was accustomed to. He was a wholesale dealer. He was counting them by the dozens." In 1935 Wheeler had been a three-month visitor at the Institute himself. But, as he recalled, "My wife wept when we left Chapel Hill; but then in 1976 she practically wept when we left Princeton for Texas."

On January 16, 1939, Bohr arrived in New York aboard the Swedish ship *Drottingholm*, carrying with him the knowledge that the nucleus had been split—nuclear fission had been discovered the month before by two German scientists, Otto Hahn and Fritz Strassman. It could well have been discovered several years earlier by Fermi in Italy or Joliot-Curie in France; it could also have been predicted by Bohr and Wheeler, using Bohr's liquid-drop model. Wheeler told me that he and Bohr had explored deformations in the drop, but not the ones most favorable for fission, and they were not trying to study what would happen if the nucleus was actually unstable. Considering what it might have meant if the Germans had gotten started on the atomic bomb in 1933 rather than in 1939, we can all be grateful fission was not discovered earlier. In actual fact, Hahn and Strassman did not understand the significance of their experimental results until they were explained to them by their former collaborator Lise Meitner. A Jewess, she had fled to Sweden with her nephew, the physicist Otto Frisch. It was Frisch who gave the name "fission" to the newly discovered process. Bohr was told about all of this just before leaving, but decided to say nothing until he could be sure that Meitner and Frisch would get proper credit for it.

Wheeler and the Fermis, who had just fled Italy, met Bohr and his party at the pier. With Bohr was his son, Eric, and his collaborator, Léon Rosenfeld. Wheeler told me that "Bohr had come, in the first instance, primarily to talk about the quantum theory of measurement and to try to convince Einstein about the quantum theory. But his whole course was perturbed, as was mine, by this fission business." The first perturbation occurred the very evening of Bohr's arrival. The Fermis took Bohr and his son to New York for the evening while Wheeler took Rosenfeld back to Princeton. Wheeler told me what happened next. "At that time I had a major part in running our Monday night journal club. It began, as a rule, at 7:30. We deliberately had too few chairs so that people would arrive promptly to get a seat. It was necessary to break up just a little before 9:30, because a lot of people wanted to go to the second show at the movie theater. We usually had three speakers, and as it was a Monday night, I got Rosenfeld to give one of the talks. He didn't realize that he was not supposed to talk about fission. He spilled the beans, and of course everybody was immensely excited about it. Bohr came to Princeton the next day, and somehow we just naturally began a collaboration on it. The first day or two Bohr dictated several paragraphs based on his idea that fission was just one more nuclear reaction. His whole idea always was to try to bring everything together in a harmonious way. Here the idea was to bring fission under the same set of outlooks that worked for other nuclear reactions. The particle—a neutron, say—goes into the nucleus, which takes it up. The nucleus loses all memory of how it was formed. The system vibrates. The excitation changes and moves around the nucleus, and finally the nucleus gets rid of it by giving out radiation, or giving out a particle, or doing this new thing—undergoing fission. The natural division here, as in so much of the rest of physics is between energy release on the one hand and the probability of energy release on the other. I can recall digging up a lot of material on the semi-empirical mass formula [an invention of German physicist C. F. von Weizsaecker which described the masses and binding energies of the nuclei] to see how we could take over some of the empirical masses to estimate the surface tension of the liquid drops representing different nuclei and then calculating the energy release for various breakups. Bohr was disappointed that the calculation did not show two peaks where the two regions for fission fragments are formed with the greatest probability."

While Bohr and Wheeler were developing the theory, experimental discoveries were being made about fission almost daily. As was typical of Bohr, he fastened on the one that seemed most paradoxical, most difficult to explain: the observation that both slow and fast neutrons would cause natural uranium to undergo fission, while neutrons of intermediate energy would not. Bohr, Wheeler told me, had the inspired idea that it was the rare isotope U^{235} that was responsible for the slow-neutron fission. In natural uranium this isotope makes up only 0.7 percent of any sample; which is nearly all U^{238}, the isotope that is responsible for the fast-neutron fission. Hence the two observed fission processes in natural uranium represented the fission of distinct isotopes. This extraordinary intuitive guess—which is what it was—led to one of the two ways of constructing an atomic bomb—namely, separating the U^{235} from the U^{238} to make an explosive mass of pure U^{235}. The other way, proposed a little later by Princeton physicist Louis Turner in an article in *Reviews of Modern Physics*, was to convert U^{238} into a new element. It was soon given the name plutonium, and it was also fissionable by slow neutrons. The idea that it was U^{235} that was responsible for slow-neutron fission seemed entirely crazy at the time to most of Wheeler's colleagues. He recalls making a bet of $18.36 against a penny with George Placzek, a well-known theoretical physicist. (The strange odds mimic the ratio of the masses of the proton and the electron.) In due course Placzek sent Wheeler a one-word telegram which read "Congratulations!"—along with a money order for a penny.

Wheeler has several vivid personal memories of this collaboration with Bohr. Near the beginning, one night at about ten they went to the library to see if they could find some synonym for the word fission. "Fission," Wheeler said, "was great as a noun, but we didn't like it as a verb. We tried 'splitting' and 'mitosis,' but they didn't seem any better, so finally we used 'fission.' " Wheeler also recalls a Fine Hall janitor for whom fission ran a distant second to keeping the offices neat. Wheeler told me that the janitor "scolded Bohr for spilling chalk on the floor. After that, when he finished working, Bohr would always pick up the rug and kick the chalk under it.

"Bohr had a very intense way of speaking. He would say, 'How can you *possibly* imagine such and such to be the case?' His eyes would be almost vertical instead of horizontal . . . almost vertical. At other times, when he was undecided, he would talk about one position with his head turned one way, and the other position with his head turned the other way. He was always ready to reconsider everything."

I asked Wheeler if he had the sense of doing something portentous when he was working on fission with Bohr. "I should have, but I didn't," Wheeler answered. "I think it was on March 16—two months to the day after Bohr had arrived in New York—that we had a meeting in Wigner's office in Fine Hall, the one that had been Einstein's before the Institute got its own quarters. There were four or five of us, including Bohr, Wigner, Leo Szilard, and myself. We talked about using fission for submarines, and for making a bomb. Bohr said that using U^{235} to make a bomb would require the resources of an entire nation. In the end, it took three.

"I was not as worked up about it as were Szilard and Wigner. I had the mistaken idea that we would stay out of the war. When I was in Denmark in 1934 and 1935 my Danish friends would listen to Hitler's Nuremberg rally speeches over the radio and ask me what the United States would do. It was my judgment, based on what had happened in World War I, that we would stay out. I felt that Germany would win any war in Europe, and that as terrible as Hitler was, there was a German culture that one could hope would come back. It is very unfortunate that I felt like that. If I had been more convinced, as Wigner and Szilard were, that we were going to get into the war, I would have pushed harder to begin making the bomb.

"I figured out that roughly a half million to a million people were being killed a month in the later stages of the war. Every month by which we could have shortened the war would have made a difference of half a million to a million lives, including the life of my own brother. If somebody could have pushed the project harder in the beginning, what a difference it would have made in the saving of lives! At that time, I thought of my work in nuclear physics as a temporary thing—an interlude before I got back to more fundamental questions."

Wheeler's return to what he regarded as fundamental physics was certainly aided by the arrival of Richard Feynman in Princeton in 1939. Feynman, who was only seven years younger than Wheeler, had been an undergraduate at MIT. He was already well known both for his original brilliance and his various and sundry capers. Feynman died in 1988, and in a moving article published in a special memorial issue of *Physics Today* Wheeler described his introduction to Feynman: " 'This chap from MIT: Look at his aptitude test ratings in mathematics and physics. Fantastic! Nobody else who's applying here at Princeton comes anywhere near so close to the absolute peak.' Someone else on the Graduate Admissions Committee broke in, 'He must be a diamond in the rough. We've never let in anyone with scores so low in history and English. But look at the practical experience he's had in chemistry and in working with friction.'

"These are not the exact words, but they convey the flavor of the committee discussion in the spring of 1939 that brought us 21-year-old Richard Phillips Feynman as a graduate student. How he ever came to be assigned to this 28-year-old assistant professor as grader in an undergraduate junior course in mechanics I will never know, but I am eternally grateful for the fortune that brought us together on more than one fascinating enterprise. As he brought those student papers back—with errors noted and helpful comments offered—there was often occasion to mention the work I was doing and the puzzlements I encountered. Discussions turned into laughter, laughter into jokes and jokes into more to-and-fro and more ideas."

Among the various intellectual capers Wheeler cites, some of which he had described in some detail when I was interviewing him, was the matter of the swastika-shaped lawn sprinkler. It shoots out four jets of water, and the reaction causes the sprinkler to go round and round. What would happen, the two young men wondered, if an identical sprinkler were to be built which *sucked in* water? Which way would the sprinkler turn? According to Wheeler, the entire physics department lined up on various sides of this pressing question. There was nothing for it except that Feynman build an apparatus to test the point. It was set up on the floor of the cyclotron laboratory. As it turned out, the apparatus exploded—spewing water and glass all over the laboratory—before the matter could be decided. Feynman was banned from the lab. He then proceeded to allow himself to be hypnotized in a demonstration that Wheeler attended at the Graduate College. Wheeler decided that Feynman was acting, and that indeed all hypnotism is playacting.

At one point, Feynman became interested in a problem involving the neurology of jellyfish. It seems that a jellyfish has a single nerve that goes entirely around it. Feynman wondered why stimulated jellyfish would not continue to vibrate indefinitely, as the nerve impulse went round and round. Jellyfish biologists had suggested that there were in fact two nerve impulses moving in opposite directions, which, upon meeting, would cancel each other and arrest the vibrations. Wheeler told me, "Somehow the boys worked out a way to cancel one of the impulses and the unfortunate jelly fish *did* vibrate hour after hour.

"Feynman used to come over to our house to work, or for supper. As my wife was cooking supper, he would explain to our two very young children how you could tell whether the contents of a can were liquid or solid by the way it would go when you tossed it up in the air.

"And," Wheeler went on, "the wonderful girl that he was engaged to, Arlene Greenbaum, would come down to Princeton from time to time for parties at the Graduate College. She would stay with us. But unfortunately she was pushing herself too hard. She was a student in art school, and to get the money she was giving music lessons. It made a terrific program for her, and she came down with this bug. Month after month the doctors failed to diagnose it. It turned out to be tuberculosis. Feynman married her, both of them knowing full well she would not live, and went with her to Los Alamos. I saw Arlene the week before she died in the hospital in Albuquerque, with all the tubes feeding oxygen into her. A wonderful girl. It was really a tragedy what happened." Feynman wrote very movingly about Arlene, especially in his second collection of essays, "*What Do You Care What Other People Think*?"

Feynman wrote his Ph.D. thesis under Wheeler's supervision. It was called "A Principle of Least Action in Quantum Mechanics" and it is one of the rare Ph.D. theses to have made a lasting impact in the field. In classical physics, given the forces acting and the initial conditions, a particle follows a uniquely determined trajectory between any two points in spacetime. This trajectory can be determined by minimizing a mathematical quantity associated with the system called the action. In the 1930s Dirac had hinted, but not made very explicit, that a similar principle might hold in the quantum theory—similar, but not identical, because in the quantum theory there are no trajectories, but rather probabilities that a particle initially located at some point in spacetime can be found at some other point in space at a later time. The business of the theory is to determine these probability amplitudes.

In Feynman's formulation the amplitudes are determined by summing the action over all possible histories between the two spacetime points in question. No matter how crazy such a history might appear from the classical point of view, it still must be included in the sum. Classical physics is recovered by letting Planck's constant go to zero in the formula for the action. This, it turns out, picks out a unique trajectory—the classical one. It is a very powerful way of looking at the quantum-mechanical formalism and is essential in the generalizations that have been made to the quantum theory of fields.

Wheeler was so taken by this work that he decided to try it out on Einstein to see if this formulation would change the old man's mind about the quantum theory. The result was predictable. Wheeler told me, "Einstein listened patiently to me for twenty minutes, and when I ended up by saying 'Doesn't this look perfectly beautiful? Doesn't this make you willing to accept the quantum theory?' he said, 'I still cannot believe that God plays dice. But maybe I have earned the right to make my mistakes." No beautiful formalism can alter the dictum of Wheeler's quantum-mechanical baseball umpire: "They ain't *nothin'* till I calls 'em." For Einstein "they" were *something* whether any one called them or not.

Feynman and Wheeler were working on several other problems, which they never got to finish because of Pearl Harbor. Once the United States had been attacked, there was no question about Wheeler's getting involved full-time in the war. Up until that time Wheeler's involvement with the nascent atomic bomb project had been peripheral. His paper with Bohr on the liquid-drop model of fission had been published in the open literature. (Ironically, the September 1, 1939, issue of the *Physical Review* in which it appeared also carried the paper by Oppenheimer and Hartland Snyder in which the notion of what Wheeler later named a black hole first appeared.) It took some time before the physicists, who were accustomed to the free exchange of scientific ideas, began to voluntarily restrain their communications on fission and related matters. But until Pearl Harbor Wheeler had ambivalent feelings about whether there would be a war and whether this country would get dragged into it.

Soon after Pearl Harbor, Wheeler went to the chairman of the Princeton physics department and said that if he were not granted a leave of absence to go to Chicago to work with Fermi on the first nuclear reactor, he would have to resign. He was granted a leave. His next problem was to find lodging in Chicago for himself and his family. All the real estate agents there told him that nothing was available. "Finally," he explained to me, "I adopted the practice of walking along a street, any street, and spotting a house I thought would be a nice place to rent, ringing the doorbell and saying, 'I am a stranger here and I'm looking for a place to rent. You don't know, by any chance, of a place in the neighborhood that's for rent? Or, by any chance, is your own place for rent?' I got nowhere—many tries. Finally, I tried one house. The lady said, 'No, I don't know any place for rent. No, this is not for rent. But, now that you mention it, I think we'd be willing to rent it.' That's where we lived—across from the International House."

Because of Wheeler's engineering background—and, very likely, his temperament—he was able from the beginning to function as a kind of bridge between the physicists working with Fermi to make the first self-sustaining, chain-reacting nuclear pile, and the professional engineers who had been brought in as consultants. For these reasons, Wheeler thinks, Arthur Compton, who ran what was cryptically called the Metallurgical Laboratory at the University of Chicago (in reality the reactor group) appointed him the liaison between the Chicago project and the Du Pont Company, which had the responsibility of constructing the first plutonium-production reactors.

On Thanksgiving Day 1942, the Fermi reactor in Chicago went critical. I asked Wheeler if he had been a witness. "No," he told me, "I was in Wilmington with the Du Pont people. I really wasn't interested in a reactor demonstration. To me, it was obvious that it would work." Wheeler had been going back and forth twice a week on the train between Chicago and Wilmington for several months, and finally, in February 1943, he decided to move his family to Wilmington.

One of the problems he was called upon to deal with in Wilmington was where to locate the first plutonium-producing reactor. Various sites were considered and ruled out—for example, Florida was ruled out since it had more thunderstorm days per year than any other place in the country. Finally the plant was sited on the Columbia River in the state of Washington, because of the isolation of the place and the quality of the water. (Pure water was absolutely essential as a cooling agent for the reactor.)

Wheeler recalls a Du Pont engineer, Roger Williams, who was responsible for many of the site decisions and also for plant safety. Wheeler told me, "Roger Williams realized how important health and safety would be. Du Pont had a long tradition of safety. I learned from the Du Pont people, many of whom had been plant managers, how in the very early days of the company they had had lots of explosions. You recall how du Pont, being a friend of Jefferson, tried, on a visit to America, to buy land in Virginia near Jefferson's. He was not allowed to. There was a law against it. Pennsylvania? No, a law against it. So the poor devil had to end up in Delaware. That's how come Du Pont, and the gunpowder works, were in Delaware. They had a lot of explosions until they learned to do things in the safest possible way."

"Du Pont," Wheeler went on, "is still famous for its safety record. The key idea was not to have safety experts that went around lecturing people on safety. That had about the effectiveness of pouring water on a duck's back. Instead, they trained various foremen on how to hold conferences—not to always talk oneself. 'Well, boys, the higher-ups tell us that we've got to do this safely. What are we going to do to make this safe?' Then he's instructed to shut up, and then someone would say, 'The last time we did something like this, I almost broke my ankle. I stumbled with a load of cement on that hose.' Then they get the idea of putting the hose under a board so he won't stumble on it. This paid off in safety, since people thought ahead about what they were going to be doing. It also paid off in efficiency. Roger Williams realized that there was going to be radioactivity, with all the hazards that were involved, so he put an enormous amount of effort into consulting experts and people in general. He also did a psychological thing. He called it health physics—not radiation physics—and that terminology is used not only in this country but all over the world."

As it turned out, the plant, which was known as the Hanford Engineering Works, was located in Richland, Washington. It employed some fifty thousand workers, mostly, Wheeler told me, "Oakies" and "Arkies." Wheeler said that when the prospective workers got off the train in Richland, they would be met by the "wolves"—who tried, in one way or another, to separate them from their money—and the "sheepherders," whose function was to get the workers to the plant intact. With a certain amount of affection, it seemed to me, Wheeler described a beer hall that featured about one murder per night. The windows on it were made sufficiently low that tear gas could be tossed in to calm any excessive exuberance. In July 1944, Wheeler moved his family to Richland.

"I was being paid by Du Pont," he told me, "as a kind of mascot to the enterprise. I was brought into this or that discussion group to answer this or that question. I had to do estimates on the greatest variety of things on the face of the earth. How much radioactivity would go downstream in the water? What kind of excursions in temperature could be expected? What was the danger of getting too much plutonium together? The heating and cooling of the plant—is it turned on or off? How much shifting of the graphite blocks will that cause? The channel for the cooling water—what can we do to widen or narrow that? How much will that affect the radioactivity?

"Of course, we were all the time keeping in touch with the people in Chicago who were doing experiments, by that time largely on the chemical processing of the water. Would the water have to be demineralized? If you look at the photographs of the Hanford plant as it was in the beginning, you'll see these gigantic towers going up and you will wonder, what in the world has that got to do with a nuclear plant? Well, they were the towers for demineralizing the water. They never had to be used. But if they did, it would have meant months of delay if they weren't ready on time."

In the meantime, Feynman had gone to Los Alamos. Among other things, he had figured out the size of an explosive mass of liquid, rather than solid, plutonium. This turned out to be smaller than the solid mass, further complicating the safety problem at Hanford. Wheeler told me that Du Pont had to make a 200-million-dollar decision: whether or not to build additional plant facilities to deal with this new safety problem. Wheeler computed that the probability of anything going wrong was so remote that this extra expense could not be justified. It was not made and nothing went wrong.

At 12:01 A.M. on September 27, 1944, the Hanford pile was allowed to start producing power at something like full strength. Many dignitaries were on hand, including Fermi. Wheeler was in the central laboratory, computing various things, when he began getting very disturbing reports. When the safety rods that moderated the reactor were pulled out, at first the level of reactivity climbed. This was expected. But then, mysteriously, it began to fall. Even with the rods completely pulled out, the pile, in Wheeler's words, "died the death." He added, "Everybody was scurrying around. There were various theories. One was that nitrogen from the air had gotten fixed in the pile and was absorbing neutrons. Another was that there was something in the water. But it had been one of my jobs to consider every possible way that things might go wrong. I was therefore very aware that a fission product, when it decays, could give rise to an isotope that could absorb neutrons. When, a few hours later, the reactivity began rising, I was sure that this was what had happened. The second nucleus had decayed into a third one which did not absorb neutrons. Outside my office there was a big chart of the nuclei. By looking at it I could see that the culprit had to be xenon-135, decaying into iodine-135. So I did some figuring on that. By then Fermi had come into my office and he accepted this explanation.

"But the real hero of the story was a Du Pont engineer named George Graves. He kept asking questions like 'Radioactivity? Who told the nucleus to be radioactive? Fission product? What in the hell are fission products?' Little by little, he began asking these pregnant questions. We had a couple of hour sessions most every day. Once he got into it, he insisted that, instead of the 1500 fuel tubes that we had planned to put in, we have a margin for error of another 500. Actually we had 2004. That took a lot of gumption, since it cost a lot of money. But thanks to his foresight it was possible to reload those extra tubes with uranium and give the pile the reactivity it needed to override the fission-product poison."

"It was quite a place," Wheeler recalled. "The mess hall in one of those plants was a huge place—big as a basketball court—table after table. Nobody was served until the table was filled up. Then the plates were brought with food and passed around. The people ate in absolute silence. It was not the fashion to talk. There was an entertainment program that was run on money from slot machines. It brought in really big names. They would get off the train in the middle of the night in this godforsaken town of Pasco and be brought to a place they had never heard of. In Hanford there had been these wonderful farms that had been run by irrigation—asparagus, fruit. But they had been taken over by the government and the trees were dying. Nobody had time during the war to look after things like that. The houses were put up in a big rush, and the tar pavement sidewalk squirted out the way toothpaste is squirted from a toothpaste tube. But in the hot sun the tar would crack and asparagus would come up through the cracks."

"I recall Walt Simon," Wheeler went on, "the plant manager in the beginning, telling me about the day that General Groves visited. I saw him at the end of that day. He told me that General Groves kept saying things like 'Simon, why isn't this town further along? Simon, why isn't this done and that done?' Simon would always be very polite and tell him the story. Finally this being so polite got under Groves's skin, and at the end of the day he said to Simon, 'Simon, what's the matter with you? Are you a man or a worm?' And Simon said, 'Well, General Groves, we have people back in Wilmington to answer questions like yours.'

"I also remember meeting Bohr in Washington—one of us from Los Alamos, Bohr, being a consumer of plutonium, and one of us from Hanford, me, being a producer. Bohr had just finished a discussion with Roosevelt about the idea of an open world. Bohr said, 'How could such a man as I talk to the president of so great a country in the midst of the greatest war in the history of the world? I simply put it to him man to man. What other way is there?' You probably read the last speech of Roosevelt, the one he didn't live to give, where he quotes Jefferson about science being a link between different countries of the world. Surely that was a fallout from Bohr's discussion with him."

Wheeler told me a remarkable fact about the Hanford plant that I had never heard before. It was the only plant in the United States that was shut down because of enemy action during the war. "How? Well," Wheeler continued, "the Japanese had an idea, which was a great idea—we kept it quiet in the newspapers so they wouldn't know how great an idea it was." They sent paper balloons with incendiary bombs, in the air across the Pacific. Some of them set forest fires in the Pacific Northwest. One of them draped itself around the power line that fed the water pumps to the pile and shut it down. That incident, as one might imagine, caused a great deal of anxiety around the project. Wheeler recalled that "there was a radar that kept up a surveillance against any possible overflights of planes. They picked up something and they checked all the airfields with a hundred miles. That particular thing turned out to be a flight of birds.

"But the news of the balloon somehow percolated around among the children at school. My small boy, and a friend, told my wife that the Japanese were landing from a balloon and that we have to do something about it. She thought it was preposterous. But they insisted that the Japanese would land from a balloon. So she called up security and, instead of laughing it off, they took it immensely seriously. Well, it was not many days after that that I was coming out from lunch with my colleagues. Someone pointed up to the sky and said, 'I think I see a balloon up there.' I looked and couldn't see it. Then someone else saw it, and then finally I saw it. It turned out to be the planet Venus in broad daylight."

Wheeler had enough contact with Los Alamos to know about the successful Trinity test of the atomic bomb on July 16, 1945. He was quite sure the war would now end. "I could see that the war would end," he told me. "I wanted to get my family back east, even though nobody else had an indication that the war would end. A few weeks before school began I took them up to Vancouver and put them on the trans-Canada train to Maine where they could spend some time with my wife's parents. On the train, my wife was in the observation car with the newspaper. She had known what was going on, since she had been in on it before we had any secrecy. There were the headlines about the atomic bomb being dropped in Japan. She was so excited. All these Canadians were sitting around the observation car showing no signs of any interest, any excitement. Finally she couldn't contain herself, so she said, 'Do you see this news?' They said, 'Well, yes.' She said, 'You see the war will be over in a few days.' They thought she was crazy."

In the fall of 1945 Wheeler returned to Princeton and to fundamental physics. He decided that cosmic radiation represented a relatively inexpensive way for the university to get into elementary-particle experimental physics, so he helped set up a cosmic-ray laboratory. An expert on the safety aspects of reactors, he did a fair amount of advising and consulting. He is one of the people responsible for having safety domes built over reactors to hold in fission products in case of an accident. Wheeler told me of a meeting that he had with his British counterparts on reactor safety in October 1949. "In our work, we had come up with the formula that a plant, for safety reasons, should be situated no closer to a city than 0.01 miles times the square root of the reactor's power level in kilowatts. It turned out, when we compared notes, the British had come up with an essentially identical formula, which shows that meteorology is the same in different countries. We went into earthquakes and fires, and the possibility of big oil tanks that would be ignited in case of trouble. After all that, we got to the question of what the chance would be of realizing the worst-case scenario. I argued that there was no way to figure it. You ought to figure, as well, the chance that somebody could come along who could turn off safety system A or B or C, someone who is so well-trusted that he gets through the security systems. I said that this was something that nobody could predict, because it depended on the climate of opinion at the time. I had collected figures on sabotage in World War I and II. I raised the question of what kind of person would do this. I said that such a person had to be completely trusted, technically very good, a loner, animated by some strange ideology. As I was speaking, Klaus Fuchs was listening to me across the table. A month later he was in jail for spying at Los Alamos.

"A few years ago, Sir Rudolf Peierls, who had been responsible for bringing Fuchs into the bomb project, told me that when he read that Fuchs was in prison he went immediately to see him. He said to Fuchs, 'There has been some terrible mistake. We've got to get proper legal counsel for you.' Fuchs said, 'No, there's no mistake. I was a spy.' Peierls said, 'How could you?' Fuchs answered, 'Well, I meant to give control of the world to the Russians.' Peierls said, 'But how *could* you?' 'But then I meant to tell *them* what was wrong with them.' "

Wheeler told me a little epilogue to this. In 1979, the year of the Einstein Centennial, Wheeler gave eighteen Einstein lectures in eight countries, involving eight crossings of the Atlantic. One of the lectures was in East Germany. "It was very strange compared to the others. The audience was restricted only to important political figures—the prime minister and the like. Among the list of notables, I noticed the name Klaus Fuchs. I asked an East German colleague if, during the coffee break, he would be willing to bring Fuchs over. When he came over I was careful to have my notebook in one hand and a coffee cup in the other. I didn't want to shake hands with him. We talked for about five minutes about the East German nuclear power program that he was busy with."

In 1949 Wheeler was awarded a Guggenheim Fellowship. He decided to use it to go to Paris so his children could learn French. But he commuted to Copenhagen every couple weeks to talk to Bohr. As it happened, this was also the time when there was great agitation in this country—much of it being carried out in secret—as to whether or not the United States should enter into a full-scale crash program to build a hydrogen bomb. Wheeler took no part in this discussion. He was, as he told me, "minding my p's and q's in Paris." But being in Europe so soon after the war gave him an acute sense of its fragility. "It seemed to me," he recalled, "like a house of cards which a wind from the east could blow down. At any rate, the phone started ringing in the *pension* where we were living in Paris. It was my friends from the Atomic Energy Commission in Washington, putting the heat on me to go to Los Alamos and take part in the H-bomb enterprise. I certainly didn't want to do it. I held off the decision until I could go to Copenhagen and talk to Bohr. I was staying at his house for two weeks. One day at breakfast I told him of the struggle I was having with my conscience about going back to Los Alamos. It was not his way to advise a person directly, but I can never forget a phrase he used. He asked, 'Do you image for one moment that Europe would now be free of Soviet control if it were not for the bomb?' That was enough to tip the scales."

Wheeler then went on, "I arrived in Los Alamos in February of 1950 and got up to speed on a lot of things because I had not worked on bomb design. I reviewed the classified literature and looked at ways to make things go. Of course, in retrospect, one can see that we, the whole community from the working days of Los Alamos to that time, had, as far as the H-bomb was concerned, been trying to get the right answer to the wrong question. We were then at Los Alamos still struggling to get the right answer to the wrong question. I said to myself, 'Look, we ought to have more people in this enterprise, and we're not going to get more people to Los Alamos in the present state of people's minds.' People were tired of war. The only way to get them would be to get them to come to Princeton for a year or two. So, with the approval of Los Alamos, and checking it out with the Princeton administration and with Oppenheimer at the Institute, we set up in Princeton."

Wheeler was able to get the use of some buildings that had been abandoned by the Rockefeller Institute for Medical Research, at what is now the Forrestal Center. Princeton astrophysicist Lyman Spitzer had agreed to help with the project, and because of his mountaineering interests had named it Project Matterhorn. As it turned out, he never did work on the hydrogen bomb because, while skiing in Aspen, he got an idea for using nuclear fusion in a controlled manner for making power. It was this device that Wheeler named the stellarator. It was the forerunner of what is now the most promising method for making fusion power, the so-called tokamak. This meant that the Matterhorn project became two Matterhorn projects, A and B, with B, under Wheeler's direction, engaging in bomb design.

Wheeler told me that at the time he wrote, phoned, or visited about a hundred and twenty scientists in an attempt to recruit them for the work. Very few senior people were interested, but he did manage to recruit a few of the younger ones. In December 1950, Stanislaw Ulam and Edward Teller, as Wheeler put it, "suddenly realized that we had all been asking the wrong questions. By asking the right questions they immediately put us on the right track. On the other hand, Teller was at absolute loggerheads with the group at Los Alamos. He felt that they did not have the people that were needed, nor the push that was needed. It was clear to me that we at Princeton could provide a very useful function with what we had. So instead of being a project to search over a long term for a solution, it became a short-term project to calculate out specific designs. We ran through burning calculations with this fuel or that fuel or the other fuel, or this dimension, that dimension, the other dimension. It was a mixture of hydrodynamics and nuclear physics—of design. Through the month of May and beyond, we were calculating night and day. We were using the IBM facilities in New York, the Univac in Philadelphia, and the facilities at the Bureau of Standards in Washington: all of them to meet the computing needs we had at that time. From today's perspective, it was ridiculous. But we were working with what we had. Sometimes those poor devils, like John Toll, who was a graduate student then, worked for thirty-six-hour stretches."

In June 1951 the General Advisory Committee to the Atomic Energy Commission met at the Institute for Advanced Study to discuss the current technical status of the project. As Wheeler was talking, John Toll rushed in with a chart he and his group had spent all night preparing. Wheeler said, "I interrupted my speaking while they raised the window from the outside so as to slide the chart through. The secretary pasted it up on the blackboard, and you could see from the chart how the burning progressed." It was this computation that convinced Oppenheimer that the hydrogen bomb was, in his indelible phrase, "technically sweet," and the bomb was built.

I asked Wheeler if it had, in his view, really been necessary. If we had not built it, what would the Russians have done? He pointed out that the Soviets exploded their hydrogen bomb nine months after ours, which must have meant, he is persuaded, that they were working on it while our debate was still in progress. Wheeler witnessed the first text of a hydrogen bomb on November 1, 1952, in the Pacific. It was the largest test we ever made (apparently larger than had been anticipated), although some of the aboveground Soviet tests were even bigger.

Part of the fallout from these tests was, very likely, Wheeler's attitude toward fallout shelters. While most of his colleagues looked on skeptically, Wheeler had one constructed near his home in Princeton. He told me that his architect had been through the bombardment of Budapest and told him that while it was not necessary to have a "fancy" shelter, some shelter was really important. Wheeler's view, not shared by most people, is that "shelters are a part of life in the last quarter of the twentieth century. My conscience bothers me all the time that I have not done something to get shelters for my children. The trouble is that they're scattered around so much—they're so mobile—it would take some doing. Our shelter in Princeton was not solely for us. It was for the neighborhood. The neighborhood knew that, and we had supplies enough."

Even before he finished with the Matterhorn project, Wheeler was anxious to get back to teaching and fundamental physics. Since 1952 Wheeler has written some half dozen books and innumerable articles on Einstein's relativity theory and its implications for gravitation and cosmology. It was in this context that, at a conference in 1967, he coined the term black hole. Prior to that, he had been using locutions like "completely collapsed gravitational object" to describe the collapsing stars first discussed by Oppenheimer and Snyder in 1939. In fact, Wheeler's first serious encounter with relativity, a subject he is so often identified with, did not begin until 1952. He was given permission to teach a course in the theory at Princeton that year. Einstein was still alive, and at the end of the course, he invited Wheeler to bring his students over to his house for tea. One of them asked Einstein what would happen to his house after his death. Einstein replied that it would never become a place of pilgrimage where the pilgrims would "come to look at the bones of the saint." In fact, the house is now occupied by one of the newest permanent members of the Institute for Advanced Study, Frank Wilczek. Wheeler's students from his relativity period, and their students and their students' students, populate the physics faculties of universities all over the country.

It had been Wheeler's intention to live out his professional life in Princeton. Throughout the years, he had had several offers to leave Princeton, but always turned them down. However, in 1976 he received an offer from the University of Texas at

Austin that was too good to turn down. It was not a matter of financial inducement—Wheeler went to Texas on half salary—but the fact that the Ashbel Smith and Roland Blumberg Professorship, which he was offered, had no formal retirement age. At Princeton, like most universities, the retirement age is seventy, and after that the professors are allowed to lecture only on special occasions. For Wheeler, the notion of life in a university without students and regular teaching was intolerable. He once said to me, "As far as I am concerned, graduate students are the number one way of advancing. How can you fight any battle except alongside your graduate students?" Part of his arrangement with Texas included money for students. During the first four years he was in Austin, Wheeler organized conferences for undergraduates from the United States, Canada, and Mexico—something that he called "recruiting through resonance"—and persuaded some outstanding students to come to Austin. When I visited Wheeler in Texas in 1985 he told me that he had picked a definite date to retire, but that he had only recently notified the administration. That has now come to pass, and the Wheelers have moved back to Princeton.

For a physicist of his eminence, Wheeler has had relatively little involvement in government affairs. His abiding interest has always been how nature works—how the laws of nature are put together on the deepest level. This, of course, has philosophical and metaphysical overtones, and therefore many of Wheeler's articles for the general public often sound philosophical and metaphysical. (Freeman Dyson once commented, after reading something of Wheeler's, that it sounded like *Beowulf*.) Despite Wheeler's intentions, this has given some aid and comfort to the New Age quantum-Buddhists. For example, in *The Dancing Wu Li Masters* Gary Zukav quotes the following from Wheeler: "May the universe in some strange sense be 'brought into being' by the participation of those who participate? . . . The vital act is the act of participation. 'Participator' is the incontrovertible new concept given by quantum mechanics. It strikes down the term 'observer' of classical theory, the man who stands safely behind the thick glass wall and watches what goes on without taking part. It can't be done, quantum mechanics says." Of this somewhat Beowulfian passage Zukav comments, "The languages of eastern mystics and western physicists are becoming very similar." (*Vide supra* Bell and the Dalai Lama.)

Wheeler told me that he has avoided reading certain books on the interpretation of quantum mechanics—he did not specify which—because "then I can avoid having to comment on them—anything I said would be used against me. If I damned them it would be prejudicial. If I praised them . . . trouble." Nonetheless, a few years ago he became extremely annoyed when, at a meeting of the American Association for the Advancement of Science, he was put on the same program with three parapsychologists. He prepared an uncharacteristically acidic written statement, which he handed out at the meeting. It contained, in reference to parapsychology, the phrase "Where there's smoke, there's smoke."

Both from his course at Columbia and from conversations with him, I concluded that Wheeler's views on quantum mechanics were pretty much orthodox Bohr. When I asked Wheeler if that was true, he replied, somewhat elliptically, "Yes, if you are willing to say that following Bohr is to wake up every morning ready to change your views completely." From Wheeler's point of view, at least on the morning that he described it to me, Bell's inequality is simply a part of ordinary quantum mechanics with no mystical overtones whatsoever. Wheeler has no interest in theories with hidden variables. He said to me, "We had Bohm here. I invited him to Austin for a week so he could talk about it. I don't think anyone was following the Pied Piper. In my course at Columbia I was just trying to state what quantum mechanics is as clearly as possible. That phrase 'No elementary quantum phenomenon is a phenomenon until it is a registered phenomenon, that is to say brought to a close by an irreversible act of amplification'—that is the essence of it. However, I am still puzzled by elementary quantum phenomena that are not put to use: the flash on a zinc-sulfide screen on some faraway planet where there is no life. It's just part of the grand mass of collisions between electrons and atoms that go on all the time—all over the place. It doesn't rate any special claim. There's no heavenly choir that's clapping its hands at having that happen. I am just driven crazy by that question. I confess that sometimes I do take 100 percent seriously the idea that the world is a figment of the imagination and, other time, that the world does exist out there independent of us. However, I subscribe wholeheartedly to those words of Leibniz, 'This world may be a phantasm and existence may be merely a dream, but,' he went on to say, 'this dream or phantasm to me is real enough if using reason well we are never deceived by it.' "

For a while Wheeler turned away from elementary-particle physics. He once told me, "I must have got the feeling that the only thing that could be worse than a tunnel that comes to a dead end is a tunnel that goes on forever." As a thought experiment, he invented an object which he originally called a *kugelblitz*—a ball of radiation—held together by it own gravitational attraction. Later he changed its name to a geon—the name that has stuck. As he described it to me, "It's something in which light goes around in a circle. The light energy generates an equivalent mass, and that mass holds the thing together gravitationally as it goes around in that circle. It's stable against losing any individual photon, but it's unstable in the sense that a pencil is unstable. A pencil standing on its tip can fall over one way or the other, although every individual atom in the pencil is stably held there, just as the individual photons are held together in this geon. But the geon is unstable because if you rotate it more rapidly it will fly apart into free photons, and if you rotate it more slowly it will contract inward and collapse." The geon was meant to serve as a kind of "toy model" of a black hole. "Then," Wheeler went on, "I realized that although I thought I was getting away from the problems of particle physics and the constitution of matter, I wasn't at all. Because as this thing collapses, the radiation gets denser and denser and its wavelength gets shorter and shorter. You begin to produce pairs of positive and negative electrons, and then pairs of all the other particles. So you might as well have started with the grand mess to begin with. So that didn't offer any true way through."

"After that," Wheeler continued, "I got deeply concerned with what the quantum theory has to say about this whole subject. I read all the papers on quantizing general relativity, written by wonderful people. But they certainly were not transparent to me, not in a way that one could see at a glance what was going on. At any rate, I ended up with this picture of what quantizing general relativity means—the idea of superspace. This is a space in which each point represents a three-dimensional geometry; it's a space in which a history of a geometry is not a line running through it but a leaf cutting through it. But this leaf has only one-third the dimensionality of the superspace as a whole. The three-dimensional geometries that lie on a leaf can be fitted together like the boxes in a Chinese puzzle. They can be fitted together to make the four-dimensional spacetime geometry, fitted together in a unique way."

When I told Wheeler I was not sure that I could see all the leaves fitting together, he laughed and went on, "That is only the classical story. But the quantum story is that there is no longer a sharp distinction between the three geometries that constitute most of superspace and the particular three geometries that constitute some particular spacetime. Three geometries that occur with appreciable probability are far more numerous than can be fitted together in any one spacetime. The very idea of spacetime is then a wrong idea, and with that idea failing, the idea of 'before' and 'after' also fails. That can be said so simply, and yet it is so hard for the lesson to grab hold in the world. It's a straight analogy with what you have in the case of a particle moving in ordinary spacetime. In classical physics it moves in a world line, a one-dimensional world line. But in quantum mechanics this becomes a wave packet localized near where the classical world line would have been. We know that the idea of a classical world line is, strictly speaking, absolutely wrong. Yet, for many purposes it is just the right thing to use, even though the true story would be to use a wave packet. In the same way, the idea of spacetime is absolutely wrong, but in most cases it is just the right thing to use. I wrote down the wave equation that describes the propagation of a wave in this superspace in a schematic form that was later spelled out in more detail by Bryce DeWitt. This is the equation that people like Jim Hartle and Stephen Hawking use to discuss cosmology."

"But," said Wheeler, "we're not content merely to utter words about the universe. We want to see the machinery taken apart. And, by George, we're so far from taking the machinery apart—just unbelievably far. For example, we feed in dimensions—4 dimensions, 6 dimensions, 10 dimensions, 11 dimensions—as if at the bottom of things there is any dimensionality at all. There can't be. It must be derived from some point of view, some way of doing physics, some overall picture in which dimensionality, at least in the beginning, does not even start to enter. Time. Why is there any such thing as time? Why should it be one dimensional? Time cannot be fundamental. The ideas of before and after fail at very small distances. They fail at the Big Bang. Why is the quantum there? If you were the Lord, building the universe, what would convince you we couldn't make a go of it without the quantum? In our work in physics, we take the quantum theory as a given. It's a sausage grinder. We drop our problems in, and turn the crank, and get out the answers. Where did the sausage grinder come from?"

Wheeler smiled and concluded, "It's such an exciting thing to be in physics. I must say that every day, when I wake up, it's a miracle. Why should there be any world here? It's incredible. What's the explanation for it? It must be absolutely beautiful—absolutely simple. And how stupid we are not to see it. I feel in desperation trying to find a clue all the time."

EPILOGUE

The principal conversations with Wheeler on which this account is based took place in the spring of 1985 in Austin, Texas. At the time, Wheeler, who was then seventy-four, told me that he had given the university administration a firm date for his retirement from the University of Texas, but that he had not yet made it public. He had in fact decided to retire the following year. As it happened, that year Wheeler had a triple bypass operation and then moved back to Princeton. Not knowing how his life had been affected by all of this, I was a little hesitant about troubling him with yet another interview. However, mutual friends encouraged me to call him at his present office in the physics department of Princeton University. When I did so, he invited me to come down and see him that very week. So on a lovely, sunny winter's day at the end of February I took the train from New York to Princeton and found Wheeler in his third-floor office in Jadwin Hall.

Wheeler looked totally unchanged, as did the state of his office: a warren of books, preprints, notebooks, and schedules. He said that the day before, a TV crew from "Nova" had been there to talk to him about quantum mechanics, and the next day a Japanese television crew would be by to ask him about the Bohr-Einstein debate. He showed me an early copy of his new book *A Journey into Gravity and Spacetime*, which, judging from the delightful illustrations I saw, will be one of those festive, Wheelerian intellectual banquets. Just before my visit Wheeler had sent me a copy of his latest paper, entitled "Information, Physics, Quantum: The Search for Links." It is pure Wheeler at his most Beowulfian. It features "three *questions*," "four *no's*," and "five *clues*." (The italics are Wheeler's.) Among the questions one finds "How come the quantum?" and among the clues one finds "No question? No answer!"—an allusion to the quantum-mechanical umpire ("They ain't nothin' till I calls 'em"). Along with the paper came a postcard with Wheeler's return address. It had several entries on the back, with blank spaces for responses. Among the entries were "The idea in it (if any!) which in your estimation has the best chance to move us ahead? Page no. or phrase" and "The idea most likely to be mistaken?" It reminded me of Wheeler's Columbia University course.

It was clear that we were not going to get any peace in Wheeler's office, so we repaired to a temporarily unoccupied lounge on the floor below. When I told Wheeler how well I thought he looked, he said that because of his heart surgery he was "good for another hundred thousand miles." Of his move to Princeton Wheeler commented, "Janette and I finally realized that we had been very poor parents. We had not brought up a single one of our children right; not one of them had settled in Texas. They all live up and down the Metroliner route; so we decided to move back to these parts. I had already retired from Texas. I thought that at the age of seventy-five it was not respectable to keep a chair away from some younger person. We moved into a retirement community some ten miles from Princeton, so I don't have to mow the lawn and I don't have to repair the roof. My friends here at the university were kind enough to give me an office, so we are now back in business." For Wheeler, "business" inevitably includes students. He has three Ph.D. students in Texas, which he still visits for several weeks a year, and he also supervises some undergraduate honors theses at Princeton. "Undergraduate theses," Wheeler commented, "are often more adventurous than Ph.D. theses. Students don't have to go through the whole rigamarole. I have the feeling that it is so important to learn—to teach. We both know of the little old lady who said, 'How can I know what I think, until I hear what I say?' I leave, in order to catch my bus, at 3:15 in the afternoon, before the departmental tea. The tea is where you meet graduate students, so I don't know many here. I love that phrase of Oppenheimer's, 'Tea is where we explain to each other what we don't understand.' "

I told Wheeler that I had had a number of conversations with Bell about quantum theory. "He's a wonderful fellow," Wheeler noted. "Did he say to you," Wheeler asked, laughing, " 'I'd rather be clear and wrong, than foggy and right'?" I told Wheeler that Bell had not used exactly those words, but that it certainly sounded like him. I also told Wheeler that from the time that Bell began to study the quantum theory, he had conceptual problems with it, and that I had asked Bell if, at that time, he thought that the theory might simply be wrong—to which Bell had answered, "I hesitated to think it might be wrong, but I *knew* that it was rotten." At this, Wheeler burst into a marvelous peal of laughter. The idea of the young Bell rebelling against the "rottenness" of the quantum theory struck Wheeler as incredibly funny.

I explained to Wheeler that Bell's problems were not in the mathematics but in the meaning of concepts such as "irreversible," "apparatus," "measurement," and the like, and I asked Wheeler if he had had similar misgivings when, as a teenager, he began studying Hermann Weyl's book in that field in Vermont, surrounded by cows. "No," Wheeler answered, "I had the feeling that the stuff was beautiful. I learned it from Weyl, and Weyl had the art of putting things in a lovely perspective. More so than anybody else I have ever read. That book was just a treat. So the feeling of 'rotten' would be the absolutely last feeling I would ever have about it. 'Beautiful' is what I would call it. To me it's the magic way to do it. I think that having started early and having used it in lots of different contexts, all the way from my doctor's thesis on the dispersion and absorption of light in a helium atom, to nuclear physics, to the decay of elementary particles, I feel absolutely at home with it. But John Bell's question I certainly sympathize with. An 'irreversible act of amplification'? As Eugene Wigner always says, 'What means it "irreversible"?"

I asked Wheeler if he has an answer to Wigner's question "No, I don't," he replied. "I think it is just wonderful to have puzzles like that staring us in the face. You'd be amused," he went on. "Every day I try to write down something in my notebook, although I don't always succeed, pushing things ahead just a little bit. I only got in two or three sentences this morning.

'*Nada*. The photon doesn't exist in the atom. It doesn't exist in the photodetector after the act of emission, and you have no right to talk of what it's doing in between. Nada—it's nothing.' Then there's the irreversible act of amplification where you've got a whole lot of things. It's *nada* to *nada*."

I asked Wheeler if this was not just the point of Einstein's unease. How can it be, he asked, that the photon is not there even "in between"? Wheeler agreed, and then I asked him if *he* had ever been troubled by that. "No," he said, "quantum theory does not trouble me *at all*. It is just the way the world works. What eats me, gets me, drives me, pushes me, is to understand *how* it got that way. What is the deeper foundation underneath it? Where does it come from? So that we won't see it as something that is unwelcome by friends that we admire—John Bell and many others—it will be something that will make you say, 'It couldn't have been otherwise.' We haven't gotten to that stage yet, and until we do, we have not met the challenge that is right there. I continue to say that the quantum is the crack in the armor that covers the secret of existence. To me it's a marvelous stimulus, hope, and driving force. And yet I am afraid that just the word—'hope'—is what does not eat, or possess, or drive so many of our colleagues in the field. They're content to take the theory for granted, rather than to find out where it comes from. But you would hardly feel the drive to find out wherefrom if you don't feel that the theory is utterly right. I have been brought up from 'childhood' to feel that it *is* utterly right. Here I was, reading that book of Weyl's at the age of eighteen and just crazy about it."

I then commented to Wheeler that from his perspective, the EPR experiment and all the fuss surrounding it must be something of a "nonstarter." "Yes, yes," Wheeler agreed. "Aage Bohr [Niels Bohr's Nobel Prize–winning physicist son] expressed the same point to me the last time I talked to him. He just couldn't understand why there were all these conferences—conference after conference—on EPR, when it is just the way it works. It is just exactly the wrong thing to be asking about. There is a conference coming up in Finland in August with some people I'd love to talk to, but I've written them to say that I'm not going. If you keep trying to pull apples off the apple tree, after a while it doesn't do. I hope that I am not being too propagandistic in speaking of the idea that when we see it all it will be so simple we'll all say, 'How stupid we've been all this time!' We've got to look for the right word, the right image. So you try one word for a day, for a week, for a month, or for a year, and then you give it up and try another one."

I asked Wheeler if he had ever been tempted by the hidden variables. "It's so interesting," Wheeler responded. "You've probably seen this picture of the potential that you have to use in the Bohm–de Broglie approach to describe, by hidden variables, the double-slit experiment. You see the electron coming in and doing this crazy thing. You may think that the Himalayas are wonderful, but this potential beats them by far. Yet they never tell you where the 'screwdriver' is—where in that morass of valleys and peaks the electron is going to start off. It just transforms the problem that eats them, back to square one. But, in the course of it, it encumbers the landscape with a lot of decoration."

I was curious whether Wheeler had been surprised by Einstein's reaction when, in the 1940s, he went to see him with Feynman's new formulation of the quantum theory. So I asked him. "Yes, yes, I was," Wheeler answered. "I thought it would ring a bell. Maybe I just didn't pound it hard enough. Maybe I was just too much in awe of him to beat on a table and shout." Since I had now read a good deal of Einstein on the quantum theory, I suggested to Wheeler that his objections went far beyond formalism. The way I put it to Wheeler was that he could have presented Einstein with a formalism that was written with a gold fountain pen and gold ink, and it wouldn't have changed his view. The thing that Einstein couldn't abide was Wheeler's quantum-mechanical umpire: "They ain't nothin' till I calls 'em." " 'I can't believe God plays dice' was Einstein's response," Wheeler added. I told him that I didn't think it was the matter of "dice" that bothered Einstein. After all, he had been one of the creators of modern statistical mechanics, which is built in a fundamental way on probabilities. I suggested to Wheeler that what bothered Einstein was "reality." He couldn't stand the idea that "they ain't nothin' till I calls 'em." He insisted that one calls them the way the *are*. He felt that there is an objective reality that would include a description of the photon between the detectors. It was not an improvement of the quantum-mechanical formalism that Einstein was after; his concern was with reality.

Speaking of the influence of Einstein, Wheeler told me a story involving the logician Kurt Gödel. Gödel was, perhaps, Einstein's closest intellectual companion at Princeton. The two of them made innumerable walks together between the town of Princeton and the Institute, talking mainly about the quantum theory. In any event, in the 1970s Wheeler, Charles Misner, and Kip Thorne wrote their book on gravitation. They were working in different cities, but for the completion of the book they were able to find an office at the Institute for a few weeks so that they could finish it. "We'd been slaving away," Wheeler recalled, "and I said, 'Why don't we take twenty minutes off and have a little break. What shall we do? How about going around and seeing Gödel?' So we went around, and knocked at his office. It was a nice spring day—sunny—but he was in there with his overcoat buttoned around him and an electric heater on the floor [constant fear of the cold was one of Gödel's many eccentricities]. I introduced my young collaborators, and said that I wondered if we could ask him what he thought of the relation between his principle of the undecidability of mathematical propositions and the indeterminism principle of Bohr and Heisenberg. But Gödel changed the subject. He wanted to know whether in the course of our work on this book on gravity we had found any evidence for, or against, a preferred sense of rotation of the galaxies. We had to confess that we just hadn't even looked into that. We hadn't said anything about his theory, his exact solution of Einstein's equations of general relativity which, incidentally, I had heard him present at the Einstein celebration here some years earlier. He was disappointed with us.

"It turned out that he himself, as a preliminary step to get some evidence, had taken down the great Hubble atlas of the galaxies. Gödel, whom you think of as the mathematician among mathematicians, had taken a ruler and got the angle and made a statistics of these numbers and concluded that within the statistical error there was no preferred sense of rotation. Incidentally, I ran into a man at the Institute a couple of years ago who was working on a biography of Gödel. He had gone through the papers of Gödel, and here were these pages after pages after pages of those numbers. It took a long time for him to figure out what they were. Of course they were just this statistical work.

"About a year after our visit to Gödel I was down the hall here in the office of Jim Peebles [a prominent Princeton astrophysicist] talking to him about cosmology. Suddenly the door burst open, and a student came in and threw down on the table a big thing. 'Here it is, Professor Peebles!' So I said to him, 'What is it? He said, 'It's my thesis.' 'What's it about?' 'It's about whether there is any preferred sense of rotation in the galaxies.' 'How marvelous,' I said, 'Gödel will be *so* pleased.' 'Who is Gödel?' 'Well,' I said, 'if you called him the greatest logician since Aristotle you'd be downgrading him.' 'Are you kidding?' 'No, no.' 'What country does he live in?' 'Right here in Princeton,' I answered. So I picked up the phone and dialed Gödel, reached him at home, and told him about this. Pretty soon his questions got to the point I couldn't answer them. I turned it over to the student, and pretty soon it got to the point that the student couldn't answer them. He gave the phone to Peebles, and when Peebles finally hung up he said, 'My, I wish we talked to Gödel before we did the work.'

"But it was at a cocktail party about a year after this at the house of Oscar Morgenstern—only about eight or ten people including Gödel—that finally Gödel broke down and said *why* he had been unwilling to talk about the relation between indeterminism and undecidability. He had walked and talked with Einstein enough, so he didn't believe in quantum theory; he didn't believe in indeterminism."

The next part of this book deals with the correspondence of Einstein and his friend Michele Besso. Some of this correspondence has to do with the quantum theory, and in it one finds Einstein at his most persuasive—the sort of language that must have convinced Gödel.

Besso

LIKE MOST PHYSICISTS of my generation, I first encountered Einstein's 1905 paper on the special theory of relativity—"Zur Elektrodynamik bewegter Körper" ("On the Electrodynamics of Moving Bodies")—in an English translation published by Dover in 1952. The title of the Dover book was "The Principle of Relativity." It was a translation of a 1913 German reprint of the early relativity papers, and it also included Einstein's paper on the general theory of relativity, published in 1916. In 1952, I was a graduate student at Harvard, and although I was nominally in the Mathematics Department my real interest was in theoretical physics. By this time, I had read a few physics papers, in the *Physical Review* and elsewhere, and I thought I had some idea of how a conventional physics paper was written. I had been instructed that when writing such a paper one included footnotes referring to work that was related to the work being presented. The Dover edition's translation of the special-relativity paper had nine such footnotes, some of which contained references to Einstein's great contemporaries H. A. Lorentz and Max Planck. References like these were just what I expected, and in that sense Einstein's paper seemed "conventional" to me. I was shortly astonished to learn from a colleague who had read the paper in the original—in Volume 17 of *Annalen der Physik*, now a collector's item—that of the nine footnotes in the Dover book only four were in Einstein's original paper. The others had been added by the distinguished German theoretical physicist Arnold Sommerfeld, who had not bothered to differentiate them from those of Einstein. None of Einstein's contained any reference to anyone; thus it was not apparent which of the contemporary papers, if any, he had read while he was creating the special theory of relativity. The paper's lone acknowledgment occurs at the end of the text, and reads, "In conclusion I wish to say that in working at the problem here dealt with I have had the loyal assistance of my friend and colleague M. Besso, and that I am indebted to him for several valuable suggestions."

At the time I read this, I was certainly not an expert in the history of modern physics. But I had taken several courses taught by Philipp Frank, who had been Einstein's friend and colleague since soon after the publication of the special-relativity paper. The courses dealt with the history of the invention of the relativity theory and the quantum theory, and to the best of my recollection the name of Besso had never been mentioned. It was quite clear to me, as inexperienced as I was, that Besso, whoever he may have been, was not an important physicist—that is, not in a class with Planck, Lorentz, or Sommerfeld. Indeed, in the rest of the Dover volume there was not a single reference to Besso. I had read Professor Frank's biography "Einstein: His Life and Times" when it was published, in 1947, and I couldn't remember whether he had mentioned Besso in it. I looked through the book again and found only one reference, which reads:

> At Bern his chief companion . . . was an Italian engineer named Besso. He was somewhat older than Einstein, and a man of critical mind and a highly nervous temperament. He was often able to offer pertinent critical remarks on Einstein's formulations, and also responded vigorously to those ideas of Einstein's which were new and astonishing. He frequently remarked about new ideas: "If they are roses, they will bloom." Around Einstein and Besso there gathered a small group of people interested in science and philosophy, who often met to discuss such questions.

That was that.

I might have asked Professor Frank more about Besso, but I didn't. My interest was less in the history and philosophy of science than in the theory of elementary particles, which I was working night and day to understand. So I forgot about Besso until the early 1970s, when I was writing my own biographical study of Einstein. What attracted me then was a few fragments from letters that Besso had written to Einstein long after the early days in Bern—fragments I had seen quoted in historical studies by physicist Gerald Holton, and particularly in his paper titled "Mach, Einstein, and the Search for Reality." The Mach in question is the Austrian polymath Ernst Mach, whose historical polemic "The Science of Mechanics," published in 1883, was devastatingly critical of Newton's notions of "absolute" space and time, and encouraged Einstein to formulate his non-Newtonian characterization of space and time. In my book I quoted from Holton's paper and noted that Besso's first name was Michelangelo—at least, that is one version of his first name. I repeat the quote here, since it gives an enticing hint that these fragments represented the tip of a very interesting iceberg. Einstein is commenting on a manuscript sent to him which propounded Machian ideas—ideas that by this time, 1917, Einstein had long abandoned. Mach was an extreme positivist—for him, direct experience was primary—whereas Einstein, after inventing the general theory of relativity, had become more and more of a Platonic idealist, convinced by his success in relativity that pure invention would suffice to make theoretical physics. Indeed, in his view theoretical invention was often a surer guide than experimental results—a position that most physicists would argue led him astray when it came to the quantum theory. The quotation from Holton goes as follows:

> Einstein commented: "He rides Mach's poor horse to exhaustion." To this Besso—the loyal Machist—responds of 5 May 1917: "As to Mach's little horse, we should not insult it; did it not make possible the infernal journey through the relativities? And who knows—in the case of the nasty quanta, it may also carry Don Quixote de la Einsta through it all!" Einstein's answer of 13 May 1917 is revealing: "I do not inveigh against Mach's little horse; but you know what I think about it. It cannot give birth to anything living, it can only exterminate harmful vermin."

This was the only substantial reference I made to Besso in my book. One day a few years later, I received a letter from Freeman Dyson, with whom I have corresponded regularly over the years. In his letter Dyson asked if I had read the Einstein-Besso correspondence, which had been published in its entirety by Hermann in Paris in 1972—an edition that presented the letters, in the original German, with a translation into French. He said that the letters were quite remarkable, and that, with my interest in Einstein, I should read them. By this time, I had had enough of reading about Einstein (my book had taken more than two years to write), and, besides, the fact that the letters were not translated into English was something of a disincentive. While I can read French reasonably well, I read German only with great difficulty, so I simply decided to read the letters once they were translated into English. I proceeded to forget all about Besso again, but not before noting that the Dictionary of Scientific Biography, which has an entry on almost everyone, has none on Besso.

This is where things stood until a year or so ago. At that time, I received a letter from a friend in West Orange, New Jersey—a retired physicist named Gerhard Lewin—who, apropos of something I had written about Mach, wrote to ask me if I had read the Einstein-Besso correspondence and offered to lend me his copy if I had not. Before he brought it to me, and before I got around to tackling the French—and the German, with the help of a dictionary—many months went by. I finally did finish reading the correspondence, and of all the Einstein letters I have read these are surely the most striking, on a purely human level. Einstein's letters to such contemporaries in physics as Max Born and Erwin Schrödinger are much more profound in terms of scientific content, but nothing compares with the intimacy of the Einstein-Besso correspondence. Einstein was not given to close friendships—"the merely personal," as he once put it—but these letters are filled with "the merely personal," even though the deep issues of physics and its philosophy are never very far away. The two men exchanged letters for fifty-two years. The last letter in the book—one of the most memorable, which I will discuss later—was written by Einstein to Besso's family, and it is dated

March 21, 1955, shortly after Besso died. Einstein died less than a month later, on April 18th.

The collection has been wonderfully edited by Pierre Speziali, a Swiss historian of science and professor emeritus of the University of Geneva. He reports in a preface that he first met "Michele Angelo" Besso in the autumn of 1946. Besso, who was also Swiss (*pace* Professor Frank), had been living in the countryside near Geneva since 1939. He had retired there after working for much of his life as a patent examiner in the Swiss Federal Patent Office, in Bern. Einstein, too, had worked there as a patent examiner, from 1902 to 1909. The proximity of the two men during that time accounts for the lacuna in the letters from mid-March, 1903, to November, 1909—a great pity, since if we had letters from that period we might know a good deal more about the genesis of the special theory of relativity than we do. On the other hand, we can be grateful for the fact that long-distance telecommunications were so poorly developed during most of the span of these letters. With the telephone and all the rest, one is led to wonder if correspondence like this will become a thing of the past. In any event, in 1946 Pierre Speziali had just been named an assistant to the mathematics faculty at the University of Geneva, a post whose obligations included looking after the mathematics library. Each Thursday morning, a little old man ("*un petit vieillard*") with the face of a prophet would appear in the library, no matter what the weather, and lose himself among the books. Speziali began talking to this visitor, who, needless to say, was Besso. (Speziali no longer remembers whether the discussions were in French, Italian, or German, since Besso was at least trilingual.) These discussions continued on a regular basis until Besso's death, and sometimes concerned Einstein.

In June of 1961, Speziali learned from a colleague of the existence of the letters. Although he had no notion of their content, or even of their number, he decided to try to publish them, and, to this end, contacted Besso's son, and only child, Vero. Vero had seventeen letters—six predating 1918, and the rest from the years 1950 through 1954. All were from Einstein to Besso. Vero thought that there might be others stored in the basement of his country house, in one or another of fifteen boxes that had traveled with Besso on his sojourns in various parts of Europe. Professor Speziali reports that he will never forget the days, late in September of 1962, when he and Vero went through the boxes. They discovered, among thousands of letters Besso had received from other physicists in Europe, fifty-eight letters from Einstein. The letters and the rest of the contents of the boxes were beginning to suffer water damage, and showed signs of having been nibbled by rats. Further searches turned up still more letters, and by 1968 a total of a hundred and ten letters from Einstein to Besso and a hundred and nineteen from Besso to Einstein had been discovered. Vero Besso gave Speziali permission to publish his father's letters to Einstein, and the Einstein estate—thank God—gave him permission to publish Einstein's letters to Besso, whence the collection under discussion. The fact that Speziali was teaching in Geneva no doubt accounts for the translation into French—something for which I am personally grateful.

I am also grateful to Professor Speziali for an enlightening biographical sketch of Besso and his forebears, which appears in the introduction. In it we learn that Besso was born in Riesbach, near Zurich, on May 25, 1873. (Einstein, incidentally, was born on March 14, 1879, in Ulm, Germany.) The first traces of the Besso family are found in seventeenth-century Spain, and the name Besso may be a deformation of the Spanish adjective *basso*, meaning of short stature. Besso shared this characteristic with other members of his family. The family was Jewish, and, like many Jewish families of the time, migrated to whatever country appeared to demonstrate a tolerance for Jews. To judge by the family's given names, most of the migration was to Italy, but some family members are known to have settled in Turkey. Insofar as there was a family business, it was insurance, which was the occupation of Michele's father, Giuseppe. Giuseppe was born in Trieste, but in 1865 he moved to Zurich, where he met his wife, a Swiss citizen, who was from Mantua. In 1879, he moved his family back to Trieste; he had meanwhile taken out Swiss citizenship for himself and his children—Michele was the oldest of five. (It must have been relatively easy to buy Swiss citizenship in those days; Einstein did so in 1901.)

Michele was, it appears, a precocious child. In a diary written in 1953 he recalls having "learned to read around the age of five from a book that explained physics, technology, and astronomy—almost everything one could know about these subjects, as they were ninety years ago, without the use of formulas." He was less successful in the *Gymnasium*: he and a comrade were expelled for sending a petition to the director complaining about the poor education of one of their mathematics teachers. As a result of this episode, Michele was sent to Rome to live with his uncle Beniamino and finish his high-school studies. He then spent a year at the University of Rome, and a report card discovered among his papers shows, not unexpectedly, that he was an excellent student of mathematics; because of that, he decided to continue his university work at the Swiss Federal Polytechnic School, in Zurich, which he entered in 1891. Five years later, Einstein matriculated at the same institution.

The date of Einstein's first encounter with Besso is not clear, but the circumstances are. It happened because of music, as did so many things in Einstein's social life. Einstein had discovered a family in Zurich, the Hünis, with whom he could play the violin, and at one of these soirées he met Besso. The two young men hit it off immediately and began to see a good deal of each other. On one occasion, Einstein brought along to the soirée Anna Barbara Winteler, the eldest daughter of the family with whom he and his sister, Maja, had boarded before he entered the Polytechnic. Fräulein Winteler and Besso fell in love, and they were married early in 1898. (Eventually—in 1910—Maja married the youngest Winteler son, Paul, and this gave Besso still closer ties with the Einstein family.) Vero was born that November, and the next year the young family moved to Milan, where Besso took up work as a consulting engineer; he had studied engineering at the

"Poly." In 1901, they moved to Trieste, and they remained there until 1904, when, on the strength of Einstein's recommendation, Besso was engaged as a patent examiner in Bern.

Another brief excursion into the Besso family history reveals a good deal about both men. A sister of Besso's, Bice Margherita Luisa Besso, married a Florentine count by the name of Rusconi. One of the Rusconi daughters, Laura, married the writer Niccolò Tucci. In 1947, Tucci accompanied his mother-in-law on a visit to Princeton to see Einstein. He wrote an article about the visit, titled "The Great Foreigner," which appeared in *The New Yorker*. In one passage Tucci describes a conversation between Bice and Einstein:

> "Herr Professor," she asked, in German (the whole conversation, in fact, was in German), "this I really meant to ask you for a long time—why hasn't Michele made some important discovery in mathematics?"
>
> "*Aber*, Frau Bice," said Einstein, laughing, "this is a very good sign. Michele is a humanist, a universal spirit, too interested in too many things to become a monomaniac. Only a monomaniac gets what we commonly refer to as *results*."

When it came to physics, Einstein was certainly a "monomaniac." He could stubbornly contemplate a problem for decades. But he accepted in Besso, with great affection, what many people considered his weakness—a certain inability to focus. Besso's letters to him often dealt almost exclusively with science—much of it bizarre—and part of the fascination of reading the correspondence is watching Einstein sort his friend out. Some of the issues involved such deep matters as why there is aging if all the underlying physical processes—the submicroscopic ones—seem to be perfectly reversible in time. (This is still the subject of active research.) Besso thought that the answer might have something to do with quantum theory, while Einstein maintained (correctly, I think) that quantum theory shed no light on the matter. It has to do, rather, with the way systems with large numbers of components evolve from less likely to more likely configurations; that is, entropy increases. One interesting feature of the letters is that as Besso got older the science in his letters got increasingly befuddled, which provoked Einstein to explain himself in ever greater detail. Toward the end of their lives, Einstein was writing him long letters that are veritable tracts on science and its philosophy.

The first letter in the collection, written by Einstein in January 1903 and sent from Bern to Trieste, is typical of the rest: a mixture of "the merely personal" and physics, often in the same paragraph. To a scientist, there is nothing strange about that; we all have colleagues to whom we write in this way. Einstein begins by describing his family life. He had, that month, married Mileva Marić, a fellow student at the Polytechnic, over the strong objections of his family. (He and Mileva had had a daughter born out of wedlock, who has disappeared. Nothing in the Besso correspondence refers to this episode, and one wonders whether Besso ever knew about it.) Eventually, the marriage went very badly, and much of the later correspondence is devoted to Besso's attempts to help out, especially with Einstein's younger son, Eduard—Tete, as he was known—who was born in 1910 and had severe psychological problems. (Tete died in 1965 in a psychiatric hospital in Switzerland. The elder son, Hans Albert, was born in 1904. He had a distinguished career as a professor of engineering at Berkeley, and he died in 1973.) But the opening lines of Einstein's first letter to Besso sound like those of a just married chauvinistic male. He writes, "I am now a married man, and my wife and I lead an extremely agreeable life. She occupies herself perfectly with everything, cooks very well, and is always cheerful." That having been dispensed with, Einstein gets down to describing his latest work on the statistical mechanics of atoms. He had reinvented, without knowing it, the statistical mechanics of people like the American physicist Willard Gibbs, gaining such a sure mastery of the discipline that he was able to apply it fearlessly two years later in his paper introducing the quantum of radiation. The letter turns to a discussion of Maja's financial problems and then reverts to science, establishing the style that continued for the next fifty-two years.

Later in 1903, Einstein arranged for Besso to work in the patent office, and in January 1904 Besso moved to Bern to take up his new post. There followed, by all accounts, five wonderful years, which both men recalled as perhaps the happiest of their lives. The two families were, more or less, inseparable; the wives got on very well. Each day, Einstein and Besso walked at least one way together between their homes and the patent office, using the occasion to discuss ideas in pure physics and philosophy. There has been some speculation as to how serious the job of patent examiner could have been, since while Einstein was thus employed he laid the foundation of much of twentieth-century physics single-handed. As far as I can tell—and the letters in which the two men recall those days seem to bear this out—it was a substantial, full-time job, which both Einstein and Besso took very seriously. Einstein got considerable pleasure from examining the inventions; he liked technological gadgets, and some years later he himself invented a new kind of refrigeration device.

What one would give to know what the two men discussed while Einstein was formulating the relativity theory! We do know that it was Besso who had introduced Einstein to Ernst Mach's writing, sometime earlier. Perhaps they discussed that. One can conjecture that the discussions must have concerned issues mainly to do with relativity, because neither of the two other masterpieces that Einstein produced in 1905—the paper on Brownian motion, in which the existence of atoms was, in a certain sense, established, and the paper on the photoelectric effect, in which the light quantum made its appearance—contained any acknowledgment of Besso (or, indeed, of anyone else). We also know that at this time Einstein did not have access to a proper technical library, or to any professional theoretical physicist. When it came to relativity, Besso appears to have been his only sounding board.

In 1909, Einstein accepted a job as an associate professor at the University of Zurich, and the letters begin again, with a postcard from Einstein in Zurich to Besso in Bern. In it Einstein makes a familiar academic complaint—that because of his teaching duties he has less free time than when he was examining patents for eight hours a day. By 1915, when—at least on a personal level—the letters take on a darker tone, Einstein had made three career moves. In March of 1911, he was called to the German University, in Prague. A year later, he went back to Zurich to become a professor at the Polytechnic. (His successor in Prague was Philipp Frank.) Finally, in 1913, he was appointed to a research professorship under the aegis of the Prussian Academy of Sciences, in Berlin. He moved there in 1914, with Mileva and their two sons, but Mileva returned to Zurich soon afterward, and their marriage was at an end. (Einstein wanted a divorce, but Mileva resisted, and they were not divorced until 1919; as part of the settlement, Einstein agreed to give her the proceeds of any Nobel Prize he might someday win. That came in 1922.)

There was also the First World War, which closed the Swiss-German frontier, cutting Einstein off from his children. Besso acted in loco parentis as best he could. Einstein suffered at this time from ill health, brought about, very likely, by the scarcity of decent food in wartime Berlin. To add to his difficulties, he had alienated most of his German colleagues by his pacifist attitude. All these things are mentioned only briefly in the letters, however, for this was the period of Einstein's greatest act of creation—that of the general theory of relativity and gravitation. The letters to Besso are bursting with enthusiasm about the new theory, which he published in March of 1916. He writes the following October that he has just returned from a visit to Leiden to see H. A. Lorentz and has found the general theory of relativity already very much alive—"*schon ganz lebendig*," in the German original. In the interstices of the physics there are continual references to Einstein's children. Tete's mental health appears to have been steadily deteriorating, and in March 1917 Einstein writes:

> The state of my younger son causes me many worries. It is not to be hoped that someday he can become a man like others. Who knows, perhaps it would have been better if he had left this world before having known life. For the first time in my life, I feel responsible, and I reproach myself.

A year later, Einstein asks Besso to buy a book for Tete, since he cannot send it from Germany, and in ensuing letters there are a few references to visiting the boys in Switzerland; however, by the 1920s they have all but disappeared from the correspondence. In 1919, Einstein married his widowed cousin Elsa Einstein Löwenthal, and became the stepfather of her two daughters, Ilse and Margot, who adopted his name.

One of the most fascinating letters in the collection, written by Besso to Einstein on January 17, 1928, may, Dr. Speziali remarks, never have been sent. It certainly did not provoke a response in kind from Einstein. For some reason, Besso feels a need to sum up all the things the two men owe to their long-standing friendship. It is not clear from the letter just what his intent is, but the tone is extraordinary. Besso writes that "the last two times we saw each other, you reminded me of my doctoral thesis, which in fact was never finished." Why this should have been a concern at such a late date is unclear, although Besso had been in some trouble in his professional life and Einstein might have felt that Besso's having a Ph.D. would enhance his chances. The letter goes on:

> For a long time, I have been thinking of how many other ties, of a completely different nature, bind the two of us together. I owe to you my wife and, along with her, my son and grandson; I owe to you my job, and, with that, the tranquility of sanctuary protecting me from people, as well as the financial security for hard times. I owe to you the scientific synthesis that without such a friendship one would never have acquired—at least, not without expending all one's personal forces—and you know even better than I do what an immense sense of extrapersonal order comes

> with such knowledge. On my side, I was your public in the years 1904 and 1905; in helping you edit your communications on the quanta I deprived you of a part of your glory, but, on the other hand, I made a friend for you in Planck.

The statement about the editing—a strange one, which also puzzles Speziali—seems to imply that Besso had some sort of role in Einstein's work on the light quantum. The reference to Planck is puzzling, too. Planck introduced the idea of the quantum, as a computational device, prior to Einstein, but for years he would not accept its implications about the breakdown of classical physics—a breakdown that Einstein perceived in 1905 and presented that year in his paper on the photon. He wrote a second, and less well-known, paper in 1906, in which he made clear the fact that Planck had not fully understood the implications of his own theory. Einstein could be quite sarcastic when it came to noting the limitations of other people's work. Perhaps Besso had persuaded him to tone down an early version of the 1906 paper, which might have been caustically critical of Planck. Planck was later instrumental in getting Einstein his job in Berlin. "Furthermore," the letter continues, "it is perhaps thanks to my defense of Judaism and the Jewish family that your private life took a different turn, and that I had to take Mileva back from Berlin to Zurich." This curious statement is also unexplained. However, Mileva Marić was not Jewish, and Einstein's second wife was; one wonders if Besso was making some sort of comparison.

In none of the other letters does Besso draw up such a bill of particulars. The reference to Besso's job is also interesting. As has been noted, Einstein had procured it for him in the first place. At the end of 1908, Besso left the patent office and for the next several years earned a desultory living as a consulting engineer and university lecturer. In 1919, he was reengaged by the patent office, and he remained there until his retirement in 1938. But in 1926 he had nearly been fired by the director, one Friedrich Haller. At that point, Einstein, who was by then the most famous scientist in the world, stepped in to help his friend again. He wrote a letter in support of Besso. This letter is reproduced in the collection, and what I find noteworthy is its candor. Einstein does not attempt to disguise Besso's fallibility but argues that his strengths overcome it. He writes:

> The great strength of Besso resides in his intelligence, which is out of the ordinary, and in his endless devotion to both his moral and professional obligations; his weakness is his truly insufficient spirit of decision. This explains why his successes in life do not match up with his brilliant aptitudes and with his extraordinary scientific and technical knowledge. It is also the reason that too few dossiers in the patent office have his name on them. Everyone at the patent office knows that one can get advice from Besso on the difficult cases; he understands with extreme rapidity both the technical and the legal aspects of each patent application, and he willingly helps his colleagues to arrive at a quick disposal of the case in question, because it is he, in a manner of speaking, who provides the illumination and the other person will power or the necessary spirit of decision. But when it is up to him to settle the matter, his lack of decisiveness is a great handicap. This has resulted in a tragic situation: one of the most precious employees of the patent office, one I would qualify as irreplaceable, gives the impression of lacking efficacy.

I do not know whether Besso ever saw this letter, but it saved his job.

In 1933, Einstein left Europe for America and never returned. He settled in Princeton, and the first letter to Besso written from Princeton is dated February 16, 1936. There is a brief mention of Tete, followed by an analysis of the social and economic situation in the United States in that Depression year. He writes:

> Here things are not getting better, despite all the natural riches. I believe, however, that the Americans will find a way that will lead the individual to economic security before it is too late. One is less held back by rigid traditions and class prejudices than one

is in Europe, and people are used to the instability of property and of the situation.

By 1938, Einstein was able to write Besso that the United States genuinely pleased him. In a letter dated August 8th, he notes:

> One rarely finds anyone who would prefer to return to Europe, to that more refined Europe. I know that you have an incurable weakness for your Italy, as the German Jews have for Germany. That sort of sentimental weakness is explained by our longing to settle somewhere on this unstable earth, because we are victims of the deceptive illusion that the goyim have a country and we do not. But I think that a home where a reasonable man is condemned to silence is not one. A German jurist, a goy, who is married to a Jewess and who had a lot of difficulty making ends meet here, responded, when I asked him if he ever felt homesick, "But why should I? I am not Jewish!" That man understood.

The colloquy with the jurist is one of the few flashes of humor in the correspondence. Einstein had a wonderful sense of humor and loved jokes—especially Jewish jokes—but his exchanges with Besso seemed to come out of a different part of his psyche. That year, incidentally, Besso was converted to Christianity. He informed Einstein of this in a letter, to which there does not seem to be any answer.

In the 1940s, the correspondence became more one-sided. Einstein would receive several letters on various subjects from Besso before something—usually of a scientific or philosophical nature—provoked a long response. The replies having to do with the scientific method have been the subject of much study by historians of modern science, such as Holton. The ones on physics I find a little sad. During this period, Einstein was pursuing what he called his "*violon d'Ingres*"—his unified field theory. From our perspective, it is difficult to understand what this program was really about. It looks, at least to me, like an all but random shuffling of mathematical formulas, with no physics in view. And what a time this was for physics! The so-called strange particles were just being discovered, and the quantum theory was proving ever more powerful. Einstein simply was not much interested. His position was that it was useless to try to understand this new physics until the electron was understood. We now believe that understanding the electron is such an intimate part of the new physics that the electron cannot be understood by itself. But Besso took all his old friend's attempts extremely seriously, and Einstein gave him detailed explanations of his various formal manipulations. It was a dialogue that somehow reminds me of the plays of Samuel Beckett.

One can, of course, read these wonderful letters from many different points of view—historical, philosophical, biographical, and so on. In view of the subject matter of this book—the quantum theory—these letters, especially those from Einstein to Besso, also provide an intimate view of Einstein's evolving attitude toward the theory. The first mention of the theory, the old quantum theory, in the letters dates back to 1911 Einstein was still writing to Besso about it in the 1950s, just prior to his death. What I propose to do here is to give a sampling from these letters, quoting in some detail from Einstein's letter to Besso dated October 8, 1952. This letter contains one of the clearest and most compact descriptions of Einstein's mature point of view that I have found anywhere. This must be the same sort of language that Bohm, who saw him during this period, found so persuasive. One can also see what must have driven Bohr to distraction. The translations are mine since, as of this writing, no English translation of these letters has been published. I will, very briefly, review the context of the various letters.

As I have noted, in March 1911 Einstein accepted a position in Prague. On the thirteenth of May he wrote to Besso from Prague. He begins the letter by turning down an invitation from Besso to visit him in Switzerland, and the reason he gives is one that will be only too familiar to academics—he needs his vacation, presumably the summer vacation, to do his research in peace. He goes on to say, "My situation and my institute enchant me, but the people are very foreign to me. These are people without natural reactions, without temperament and a curious mixture of pretension and servility, without any goodwill toward their neighbors. . . ." And so on. Having gotten that off his chest, Einstein gets down to physics. He continues, "However, I am compensated by the possibility of pursuing in all tranquility my scientific reveries. The work I have just finished is not very important; I am enclosing it with this letter. At this moment, I am trying to extract from the quantum hypothesis the law of heat conduction in solid dielectrics. I don't ask anymore if quanta really exist. In the same spirit, I am not trying anymore to construct them, because I know now that my brain is incapable of advancing in that direction. But I examine very carefully the consequences of this representation in order to inform myself of the limits of its applicability. The theory of specific heats can claim a victory, since Nernst has confirmed by his experiments that everything takes place more or less as I predicted it would . . . "

By 1918, Einstein, who has by then moved to Berlin, has decided that the quantum of radiation really *does* exist. On July 29, 1918, he writes to Besso from Ahrenshoop in Pomerania, "I have reflected during an incalculable number of hours on the question of the quanta, naturally without making any real progress. But I don't doubt anymore the *reality* of the quanta of radiation, although I am still entirely alone in having this conviction . . . " [All italics in this letter and the ones that follow are in the originals.] This was still the period when Einstein's colleagues—including, it will be recalled, Planck—found his photon interpretation of the photoelectric effect to have been misguided. In May 1924, Einstein writes to Besso from Kiel in an uncharacteristically optimistic tone concerning the quanta. He reports, "Scientifically, I have been plunged almost without interruption into the problem of the quanta and I truly believe that I am on the right path—if that is really so . . . My new efforts are an attempt to reconcile the quanta and the field of Maxwell. [I have not been able to figure out just what Einstein is referring to here. From our present point of view, this reconciliation is achieved in quantum electrodynamics.] From an experimental point of view, in the past few years, there are only those of Stern and of Gerlach, as well as that of Compton (the scattering of the radiations of Roentgen [x rays] with a change of frequency), that are important; the first proves the unique existence of quantum states, and the last, the reality of the momenta of light quanta." By the end of 1925 the first papers on the quantum theory are starting to appear, and Einstein writes to his old friend, from Berlin, to alert him. He says, "The most interesting thing that theorists have come up with recently is the theory of Heisenberg-Born-Jordan of quantum states. It is a veritable calculation by magic where infinite determinants (matrices) appear in the place of Cartesian coordinates. It is eminently ingenious, and sufficiently protected by a great complexity against any proof of its falsity." Then came the papers by Schrödinger, of which in May 1926 Einstein writes, "Schrödinger has brought out two excellent works on the quantum rules . . . which give one a foreboding of profound truths. Ask him to explain them to you."

The next mention of the quanta in the correspondence, which has become much less frequent on Einstein's side, is dated August 1938. By this time, Einstein had settled in Princeton and had settled his views on the quantum theory. He writes to Besso, "From the scientific point of view, I am traversing a very interesting period. You know well that I have never believed in the essential statistical foundations of physics, despite the success of the quantum theory. Now, this year, after twenty years of vain research, I have found a promising theory of fields which is a natural follow-on to the relativistic theory of gravitation. It is along the lines of Kaluza's theory of the electric field." This theory, like the ones that followed it, did not live up to its promise.

The two most complete statements in the correspondence of Einstein's final position on the quantum theory are found in two letters Einstein wrote to Besso with a month of each other in the fall of 1952. By this time, the correspondence had become still more one-sided—Besso to Einstein—and it is some measure of Einstein's feelings about the subject that he wrote two such letters to Besso within less than a month. He was also writing equally detailed letters to other old friends, such as Max Born. The two letters to Besso are so interesting that I will quote extensively from them. The first letter is dated September 10. Part of it deals with the role of time in the theory of relativity—a fascinating subject in its own right—and the rest deals with the meaning of the Schrödinger wave function ψ in quantum mechanics. Both parts are in response to some slightly befuddled comments from Besso. I will take up only that part of the letter that discusses the wave function. The quantum theory, Einstein writes, "with Schrödinger's equation determines the propagation in time of the [wave] function ψ. But this function, however, ought not to be considered as a representation of a *real state*—even for one given value of the time. On this matter one is easily deceived, because the word 'state' is used to express that which ψ represents. That it cannot be interpreted as a 'real state' one sees already from the fact that the superposition of two functions ψ of the same system is a new function ψ. The superposition of real states makes no sense, as one sees immediately with the 'macro-systems.' "

Einstein continues, "A real state is not describable in the present quantum theory, which furnishes only an incomplete knowledge of a real state. The orthodox quantum theoreticians, in general, don't admit the notion of a real state (based on positivistic considerations). One ends up, then, in a situation that strongly resembles that of the good bishop Berkeley.

"This situation, evidently, strikes many people as being uncomfortable. But it is, until now, the only way to calculate the quantum states and their transition probabilities that agrees with experiment. I am entirely convinced that the truth is situated far from the present teachings. Who knows if my nonsymmetric general relativity isn't the correct theory. The mathematical difficulties in confronting it with experiment are, for the moment, insurmountable. However that turns out, we are as far from a truly rational theory (of the duality of light quanta and particles) as we were fifty years ago!"

Besso replied at once to this letter. His response persuaded Einstein—and now, having read it several times, also me—that he hadn't really grasped what Einstein was getting at. This provoked a still more detailed letter from Einstein, written on the eighth of October. It is this letter that, in my view, gives one of the clearest accounts of Einstein's position to be found anywhere. He writes,

> Dear Michele,
>
> Your letter of 21.IX. consists, it seems to me, of two parts that are more or less independent of each other. The object of the first part is quantum theory and physical reality. What relation is there between the "state" ("quantum state") described by a function ψ and a real deterministic situation (that we call the "real state")? Does the quantum state characterize completely (1) or only incompletely (2) a real state?
>
> One cannot respond unambiguously to this question, because each measurement represents a real uncontrollable intervention in the system (Heisenberg). The real state is not therefore something that is immediately accessible to experience, and its appreciation always rests hypothetical. (Comparable to the notion of force in classical mechanics, if one doesn't fix *a priori* the law of motion.) Therefore suppositions (1) and (2) are, in principle, both possible. A decision in favor of one of them can be taken only after an examination and confrontation of the admissibility of their consequences.
>
> I reject (1) because it obliges us to admit that there is a rigid connection between parts of the system separated from each other in space in an arbitrary way (instantaneous action at a distance, which doesn't diminish when the distance increases.) Here is the demonstration:

A system S_{12}, with a [wave] function ψ_{12}, which is known, is composed of two systems ψ_1, and ψ_2, which are very far from each other at the instant *t*. If one makes a "complete" measurement on S_1, which can be done in different ways (according to whether one measures, for example, the momenta or the coordinates), depending on the result of the measurement and the function ψ_{12}, one can determine by current quantum-theoretical methods, the function ψ_2 of the second system. *This function can assume different forms*, according to the *procedure* of measurement applied to S_1.

But this is in contradiction with (1) *if one excludes action at a distance*. Therefore the measurement on S_1 has no effect on the real state S_2, and therefore assuming (1) no effect on the quantum state of S_2 described by ψ_2.

I am thus forced to pass to the supposition (2) according to which the real state of a system is only described incompletely by the function ψ_{12}.

If one considers the method of the present quantum theory as being in principle definitive, that amounts to renouncing a complete description of real states. One could justify this renunciation if one assumes that there is no law for real states—i.e., that their description would be useless. Otherwise said, that would mean: laws don't apply to things, but only to what observation teaches us about them. (The laws that relate to the temporal succession of this partial knowledge are however entirely deterministic.)

Now, I can't accept that. I think that the statistical character of the present theory is simply conditioned by the choice of an incomplete description. . . .

Your Albert

As I read these letters, I kept hearing Bell's musical Irish voice saying, "For me, it is a pity that Einstein's idea doesn't work. The reasonable thing just doesn't work."

Besso responded poetically. On the tenth of December he wrote, "This morning the radio overwhelmed me with the powerful last phrase of the Trout Quintet. On this occasion, I thought, in an orchestra the conductor is visible—in a quartet, it is the invisible creator of the music. Invisible lines link the musicians: as the canvas of probabilities links everything, everything that is realized, all reality. The most solid lines link mother and child. Solid lines link us two . . . ”

✣

Besso died on March 15, 1955, and Einstein's last letter in the collection, written to Vero and Bice, is one of the most moving and personal of all his writings. He reveals feelings about himself that I have never seen expressed anywhere else. Einstein writes:

> Dear Vero and Dear Mme. Bice:
>
> It was truly very good of you to give me, in these so painful days, so many details concerning the death of Michele. His end was in harmony with the image of his entire life and the image of the circle of people that surrounded him. The gift of leading a harmonious life is rarely joined to such a keen intelligence, especially to the degree one found in him. But what I admired most about Michele was the fact that he was able to live so many years with one woman, not only in peace but also in constant unity, something I have lamentably failed at twice.

Einstein's second wife died in 1936, and this remarkable admission is the only hint I know of that their married life was not successful. (Besso's wife died in 1944.) The letter continues:

> Our friendship was born when I was a student in Zurich, where we met regularly at musical evenings. He, the older and wiser, was there to stimulate us. The circle of his interests appeared truly without boundaries. Nevertheless, it was his critical philosophical preoccupations that seemed most characteristic of him.
>
> Later, it was the patent office that brought us together. Our conversations as we returned from the office had an incomparable charm—it was as if human contingencies did not exist at all. However, later on we had more difficulty understanding each other when we wrote. His pen did not succeed in following his versatile mind—to such an extent that it was impossible in most cases for his correspondent to guess at what he had omitted to note.

So in quitting this strange world he has once again preceded me by a little. That doesn't mean anything. For those of us who believe in physics, this separation between past, present, and future is only an illusion, however tenacious.

In thanking you cordially, I send you my best thoughts.

Your
A. Einstein

A few weeks later, Einstein, too, had quit "this strange world."

✣ *Select Bibliography* ✣

Bell, J. S. *Speakable and Unspeakable in Quantum Mechanics*. New York: Cambridge University Press, 1987.

Bernstein, J. *The Tenth Dimension*. New York: McGraw-Hill, 1989.

Bohm, D. *Quantum Theory*. New York: Prentice Hall, 1951.

Born, M. *The Born-Einstein Letters*. New York: Walker, 1971.

D'Espagnat, B. *Conceptual Foundations of Quantum Mechanics*. 2d ed. Reading, Mass.: Addison-Wesley, 1976.

Einstein, A. *Correspondence avec Michele Besso 1903–1955*. Edited by P. Speziali. Paris: Hermann, 1979.

Kuhn, T. S. *Black-Body Theory and the Quantum Discontinuity, 1894–1912*. Chicago: University of Chicago Press, 1987.

Misner, C. W., K. Thorne, and J. A. Wheeler. *Gravitation*. San Francisco: Freeman, 1973.

Moore, W. *Schrödinger*. New York: Cambridge University Press, 1989.

Pais, A. *Subtle Is the Lord*. New York: Oxford University Press, 1982.

Sakurai, J. J. *Modern Quantum Mechanics*. Reading: Benjamin/Cummings, 1985.

Wheeler, J. A., and W. H. Zurek. *Quantum Theory and Measurement*. Princeton: Princeton University Press, 1983.

✣ *Index* ✣

accelerators: design of, 11–12; strong focusing principle, 16
aging, reversibility of physical process and, 150–51
Albert Einstein: Philosopher Scientist, 38
American Association for the Advancement of Science, 131
American Physical Society, 44
Ames, Joseph Sweetman, 103–4
amplification, quantum theory and, 137–38
Anderson, Carl, 49
angular momentum, 57–62
Annalen der Physik, 27
antimatter, development of, 49
apparatus, importance of, in quantum theory, 51–53
Ashbel Smith and Roland Blumberg Professorship, 130
Aspect, Alain, 76
astronomy, locality and, 71
atomic bomb research and testing (Los Alamos), 124
atomic clocks, gravitation and relativity theory, 43–44
Atomic Energy Commission, 126–29
Atomic Energy Research Establishment (Harwell), 15–19
atomic spectra, electron orbits and, 32
atomic theory of matter, 54

Baldwin, Lydia, 101
Baltimore Federation of Church and Synagogue Youth, 106–7
Belfast Technical High School, 13
Bell, John Stewart: Bohm's work analyzed by, 65–67, 72–73; Bohr evaluated by, 51–53; California visits of, 67–68; CERN appointment, 19–20; on Copenhagen interpretation of quantum theory, 53; early life and education, 12–15, 50–51; Eastern religions—quantum theory connection, 80–83; on Einstein's abandonment of quantum theory, 84; Einstein-Podolsky-Rosen experiment discussed, 7, 49; elementary-particle physics research, 18–19; Heisenberg evaluated by, 52–53; on hidden variables, 64–65, 68–69; impossibility proof, 73; Large Electron Project (LEP), 86–88; Maharishi Mashesh Yogi and, 83–84; marriage to Mary, 17; on correlations, 62–64; Peierls's work with, 67; personal life, 7–11; reservations about quantum theory, 8–9, 20, 84–86; strong focusing principle, 16–17; TCP theorem, 17–18; uncertainty principle discussed, 50–51; *violon d'Ingres* of, 74; on von Neumann's theorem and, 64–65; Wheeler's assessment of, 136–37
Bell, Mary: CERN career, 8–10, 19–20; early life and education, 15–16; Large Electron Project and, 86–87; marriage to John, 17; retirement of, 11; Stanford visit of, 67–68
Bell's inequality (Bell's theorem), 7, 73–77; impact of, on popular culture, vii–viii, 77–80; Wheeler's view of, 131–32
Beowulf, Wheeler's writing compared to, 130–31, 135–36
"Bertlmann's Socks and the Nature of Reality," 8; EPR experiment and, 45–46

Besso, Bice Margherita Luisa, 150
Besso, Michele: analyzes friendship with Einstein, 154–55; conversion to Christianity, 157; correspondence with Einstein, 144–62; death of, 164; early life and family history, 148–49; education of, 149; influence on Einstein's relativity research, 144; failure to understand quantum theory, 162–63; meeting with Einstein, 149–50; quantum theory evolves, in correspondence, 150–51, 158–59
Besso, Vero, 147–48
Big Bang theory, 81–82
blackbody radiation, 21–23
black hole theory, 117; geon as model of, 132–33
Bohm, David, 49; Einstein and, 158; EPR experiment analyzed by, 57, 60–62; on nonlocality, 72–73; on hidden variables, 64–65; publishes *Quantum Theory*, 53–54; reactions to papers by, 65–67; relations with Einstein, 56–57
Bohr, Aage, 138–39
Bohr, Eric, 111
Bohr, Niels: Bell's evaluation of, 51–53; on Bell's inequality, 7; Born influenced by, 38; double-slit experiment, 41–42; Einstein's relations with, 38–40, 43–45, 158; Einstein's superiority over, 84; Heisenberg influenced by, 37; interest in Eastern religions, 80; "liquid-drop model" of nucleus, 109; nuclear fission research and, 110–18; origins of quantum theory and, 4; planetary atomic model of, 28–30, 104; plutonium reactors and, 123; principle of complementarity, 42; rebuttal of EPR experiment, 46–49; Rutherford's influence on, 107; Schrödinger debates with, 55–56; *schwindlig* (dizzy) thinking and quantum theory, 20, 59; Wheeler influenced by, 107–8, 131–32
Bohr orbit, 29
Boltzmann, Ludwig, 54
Born, Max, 64, 66; Einstein correspondence with, 37, 48–49, 146, 161; Nobel Prize awarded to, 37; Pauli correspondence with, 76; Schrödinger equation, 35
Breit, Gregory, 106, 108
Brown's molecular motion, 4, 55, 152
Buddhism, quantum theory and, 81–83
Buhler-Broglin, Manfred, 86–87
Bureau of Standards, 128

"Can Quantum-Mechanical Description of Physical Reality Be Considered Complete?," 44
Carnegie Institute for Terrestrial Magnetism, 105–6
CERN (Conseil Européen pour la Recherche Nucleaire): Bell's career at, 7, 9–10; Dalai Lama visit to, 80–81; elementary-particle physics, 88–89; Large Electron Project (LEP) and, 87–88; organizational structure, 19; origins of, 10–11, 19; Proton Synchrotron at, 17; work habits at, 67
Cern Courrier, 80
chemical bonding, quantum theory and, 49
China, science and civilization in, 82
Christofilos, Nicholas, 16
Clauser, John, 75–76
Colgate, Sterling, 102
community relations, nuclear research and, 87–88
Compton, Arthur, 118, 160
computers: early developments in, 106; role of, in early atomic research, 128; solid-state (condensed-matter) theory, 27–28

Condon, E. U., 109
conjugate quantities (position and momentum), quantum theory and, 36. *See also* momentum; position
Copenhagen interpretation of quantum theory, 52–53; gravitation, 69; planetary atomic model, 28–29; Schrödinger visits, 55
correlations: spin concept and, 60–62; "twins" analogy of, 62–63
cosmic radiation, elementary-particle physics and, 124–25
Coulomb, Charles Augustin, 69
Courant, Ernest, 16
Crick, Francis, 33

Dalai Lama, 80–81
The Dancing Wu Li Masters, 6, 77–78, 130–31
Davisson, C., 14, 31
de Broglie, Louis (Prince), 33–34, 66; early research, 30–31; opposition to statistical mechanics, 38; quantum theory and, 55
de Broglie, Maurice (Prince), 30
Descartes, René, 69–70
determinism: quantum theory and, 65–66; wave theory and, 35–37
DeWitt, Bryce, 134
dice-playing analogy in quantum theory (Einstein), 37, 40, 48–49, 117, 139–40
Dictionary of Scientific Biography, 146
diffraction, 24–25
Dirac, Paul, 4, 34; quantum theory and relativity, 49–50
Discourses and Mathematical Demonstrations Concerning Two New Sciences, 69
Domash, Larry, 83–84
double-slit experiment, 40–41, 139
Du Pont Company, nuclear reactor research, 118–24
Dyson, Freeman, 15, 130, 145–46

Eastern religions: Oppenheimer and, 6; quantum theory and, 80–83; Schrödinger's study of, 33
The Economist, 5
Ehrenfest, Paul, 58, 105
Einstein, Albert: abandons quantum theory, 7, 37, 44, 54, 84–85, 159–62; Besso correspondence with, 144–62; Bohm's relations with, 56–57, 66; Bohr's relations with, 38–40, 43–45; on blackbody radiation, 23–26; on death of Besso, 164; defense work in World War II, 78; "dice-playing" analogy criticizing quantum theory, 37, 40, 48–49, 117, 139–40; divorce of Mileva, 153; double-slit experiment, 41–42; emigration to U.S., 156–57; EPR experiment and, 48–49; family and marriages, 151–54; Habicht correspondence with, 3–4, 23–24; letter of recommendation for Besso, 156; local realism and quantum theory, 73–74, 76; marriage to Elsa Lowenthal, 154; on mathematical foundations of quantum theory, 56; molecular theory of heat, 55; 1905 papers, 4, 27, 55, 143, 152–53; Olympia Academy and, 3; opposition to statistical mechanics, 38; Patent Office work, 147–48, 152–53; pilot waves (*Führungsfelder*), 34; on planetary atomic model, 29–32; in Prague, 20–21; probability amplitudes and quantum theory, 117; quanta units developed by, 23; relativity theory developed by, 153–54; solid-state (condensed-matter) theory, 27–28; "spooky actions at a distance" (*spukhafte Fernwirkungen*), 44, 62, 73; on theory vs. experimental science, 145; uncertainty principle and, 43–44; unified field theory, 110, 157–58, 160; wave equation

Einstein, Albert (*cont.*)
influenced by, 33; Wheeler's assessment of, 129, 138–41
Einstein Centennial, 125–26
Einstein, Eduard (Tete), 151, 153–54, 156–57
Einstein, Hans Albert, 151
Einstein: His Life and Times, 144
Einstein, Maja, 149
electricity, 69
electrodynamics, 68, 108
electron: atomic spectra, 32; identification of, 14; "state of," 95–96; wave mechanics and, 30–31, 34
electron microscope, quantum theory and, 31
electron-positron accumulator, 86–87, 108
electrostatics, 68, 79
elementary-particle physics, 18–19, 137–38; cosmic radiation and, 124–25; Wheeler's view of, 132–33
Elements of Electricity, 14
Emeleus, Karl, 14
entropy, 150–51
EPR (Einstein-Podolsky-Rosen) experiment, 44–45; Bell's inequality and, 75; Bertlmann's socks analogy and, 8–9; Bohm's interpretation of, 54, 57, 60–62; cultural aspects of, vii–viii; impact of, on popular culture, 77–78; impact on quantum theory, 49–50; locality and, 72–73; Wheeler's assessment of, 138–39
ether, early physics research and, 68–69
Ewald, Peter Paul, 14
excited states, planetary atomic model, 29

fallout shelters, 128–29
Feinberg, Gerald, 71–72, 93
Fermi, Enrico: on nuclear fission, 110–11, 118; nuclear reactor research, 121–22; on quantum theory, 49
Fermi distance, 64–65
Feynman, Richard: atomic bomb project, 121; death of wife (Arlene Greenbaum), 116; Ph.D. thesis, 116–17; popular culture influenced by, 5; reservations about quantum theory, 77; Wheeler's influence on, 97–98, 107, 114–15
Forrestal Center, 127
four-dimensional spacetime geometry, 133–34
Frank, Phillip, 20, 33, 144, 153
Freedman, Stuart, 76
Frisch, Otto, 110–11
Fuchs, Klaus, 125–26

Galileo, 3, 69–70
Gamow, George, 55–56
Gell-Mann, Murray, 57
genetic determinism, quantum theory and, 63–64
geon, development of, 132–33
Gerlach, Walter, 60–61, 160
German University (Prague), 153
Germer, L., 31
Gibbs, Willard, 151
Gödel, Kurt, 140–41
Goudsmit, Samuel, 58–59
Graves, George, 122
gravitation: early research on, 68–69; relativity theory and, 43–44
Greenbaum, Arlene, 116
ground states, planetary atomic model, 29
Groves, General, 122–23
Gruppentheorie und Quantenmechanik, 104

Habicht, Conrad, Einstein correspondence with, 3–4, 23–24
Hafstead, Larry, 105–6
Hagner, Janette, 107

Hahn, Otto, 110–11
Haller, Friedrich, 155
Hanford Engineering Works, 119–21
Hapgood, 5
Harrere, Heinrich, 80
Hartle, Jim, 134
Hawking, Stephen, 134
heat absorption, solid-state (condensed-matter) theory, 27–28
Heisenberg, Werner: Bell's theorem and, 7; Einstein nominates to Nobel Prize, 44; matrix mechanics, 32; Nobel Prize awarded to, 37; nuclear dimensions of quantum theory, 49; origins of quantum theory and, 4; planetary atomic model, 31–32; Schrödinger and, 55. *See also* uncertainty principle
Herzfeld, K. F., 104–5
hidden variables, 54–57: Bell's analysis of, 64–65, 85; critical reactions to, 66–68; spin concept and, 62; Wheeler's assessment of, 131–32, 139–40
holistic medicine, quantum theory and, 79
Holton, Gerald, 145
Holt, Richard, 75–76
Horne, Michael, 75–76
House Un-American Activities Committee, 56
Huygens, Christian, 24
hydrogen bomb research, 126–29

I Am News, 6
IBM, 128
IMB (Irvine, Michigan, Brookhaven) detector, 88
"Information, Physics, Quantum: The Search for Links," 135–36
Ingenious Mechanisms and Mechanical Devices, 97
Institute for Advanced Study (Dublin), 33
Institute for Advanced Study (Princeton), 39, 110, 127–28
Institute for the Study of Twins, 63
interference, in quantum theory, double-slit experiment, 40–41

Jauch, Josef, 67–68
Jeans, James, 23
Johns Hopkins University, 103–4
Joliot-Curie, 110
A Journey into Gravity and Spacetime, 135
Judaism, Einstein and, 155–57

Kennedy, John F., assassination of, 67
Kirchoff, Gustav, 22
Kronig, Ralph, 58
Kuhn, Thomas, 23, 27

Lamston, Alison Wheeler, 109
Landau, Lev, 56
Langevin, Paul, 31
Large Electron Project (LEP), 10–11, 86–89
Lewin, Gerhard, 146
light: Galileo's research on, 69–70; particle theory of, 24–25; principle of complementarity and, 42; properties of, 23–24; speed of, 69–71; wave propagation, 75; wave theory of, 24–25
liquid-drop atomic model, 109; nuclear fission and, 110–12, 117–18
Livingston, M. Stanley, 16
locality concept, 68–69, 71
Lorentz, Hendrik Antoon, 31, 58, 102, 143, 153
Löwenthal, Elsa Einstein, 154
lower-energy proton-antiproton collider, 88–89
Lüders, Gerhard, 18
Lüders-Pauli theorem, 18

"Mach, Einstein, and the Search for Reality," 145

Mach, Ernst, 145, 152
magnetism, 69; spin concept and, 59–62
Maharishi Mahesh Yogi, quantum theory and, 83–84
Manchester University, 28
Mandl, Franz, 65
Marić, Mileva, 151, 153, 155
Mathematical Foundations of Quantum Theory, 56
matrix mechanics, 32
Matthews, Paul, 17
Mauchley, John, 106
Maxwell, James Clerk, 25, 54, 68
McConnell, R. A., 79
Mechanics, Molecular Physics, Heat and Sound, 14
Meitner, Lise, 110–11
Mermin, David, 78
meson, discovery of, 108
Metallurgical Laboratory (University of Chicago), 118
Misner, Charles, 140–41
Moke, Verdet, 101
molecular theory of heat, 55
Møller, Christian, 108–9
momentum: EPR experiment and, 46–48; quantum mechanics and, 36; spin concept and, 59–60. *See also* angular momentum
Morgenstern, Oscar, 141
Morton Salt Company, 88
Murray, Bob, 107
My View of the World, 80

National Research Council Fellowships, 105–7
Natural Philosophy of Cause and Chance, 64
Nature, 47
nature, laws of, 130–31
Needham, Joseph, 82
Nernst, Walther, 28, 159
neutrinos, 88; Z^0 concept, 90–91
neutrons, slow and fast, in nuclear fission, 112
New Age mysticism: influence of quantum theory on, 6; quantum theory and, vii–viii, 130–31
Newton, Isaac: gravitational theory, 68–69; laws of motion, 13; on light, 24; space and time concepts, 145
Nobel Prize in Physics, 37–38, 44
nonlocality: defense applications, 78; parapsychology and, 79
nuclear fission, quantum theory and, 110–18
nuclear power plants: early experiments at Hanford, 120–22; enemy action against, 123–24; Wheeler's research on, 118–24
nucleus: discovery of, 28; liquid-drop model of, 109; slow neutron absorption by, 108–9

Olympia Academy, 3
"On the Einstein-Podolsky-Rosen Paradox," 7
"On the Electrodynamics of Moving Bodies," 143
Oppenheimer, Robert: black hole research, 117–18; Bohm and, 56; collapsing star research, 129; hydrogen bomb research, 127; on electrodynamics, 108; Reith lectures, 5, 38–40; studies in Europe, 104; study of Eastern religions, 6; Wheeler's relations with, 106
Optiks, 24

Pais, Abraham, 38–39, 42
parapsychology, quantum theory and, 79–80, 131
parity violation in weak interactions, 106
particle theory of light, 24–25; principle of complementarity and, 42

Pauling, Linus, 49
Pauli, Wolfgang, 18, 76; Heisenberg and, 52–53; interest in Eastern religions, 80; origins of quantum theory and, 4; spin concept, 58
Peebles, Jim, 141
Peierls, Rudolf (Sir), 15, 17, 67, 125
philosophy and physics, 13, 94–95, 130–31. *See also* Eastern religions; reality
photoelectric effect, 152, 159
photon polarization, 75
Physical Review, 44, 64–65, 117–18, 143; Bohr rebuttal of EPR experiment, 47; importance of, 74–75
Physical Review Letters, 75–76
Physics magazine, Bell's inequality published in, 74
Physics Today, 114
Picasso, Emilio, 86–87
Placzek, George, 112
Planck, Erwin, 21
Planck, Max, 21, 143; atomic theory of matter, 54–55; Besso introduces Einstein to, 155; blackbody radiation theory, 27–28; opposition to statistical mechanics, 38; photoelectric effect, 159; radiation formula, 22–23, 27
Planck's constant: conjugate quantities and, 36; probability amplitudes, 117; Z^0 principle, 90–91
planetary atomic model, 28–38, 50; angular momentum and, 58–62
plutonium: liquid vs. solid, 121; nuclear fission and, 112; nuclear reactors and, 118–24
Podolsky, Boris, 44–45
popular culture, quantum theory and, vii–viii, 1, 5–7, 77–81
position: EPR experiment and, 46–48; quantum mechanics and, 36; spin concept and, 59–60
positrons, discovery of, 49
principle of complementarity: EPR experiment and, 48–49; spin concept and, 61
"A Principle of Least Action in Quantum Mechanics," 116
probability: amplitudes, 116–17; hidden variables and, 64; statistical mechanics, 140; wave function and, 35
Problems of Modern Physics, 102
Project Matterhorn, 127–29
"Proposed Experiment to Test Local Hidden Variables," 75–76
proton, 65
Prussian Academy of Sciences, 153

quanta: Besso-Einstein correspondence on, 159; blackbody radiation, 23; double-slit experiment, 41; polarization, 75
Quantum Mechanics, 49–50
Quantum Theory, 49, 53–54
Quantum Theory and Measurement, 9
quantum theory (quantum mechanics): absurdity of, 4, 6; Bell's inequality and, 74–77; Bohm's interpretation of, 53–55; chemical bonding and, 49; criticisms of, 20–21; Eastern religions and, 80–83; Einstein and, 158–62; genetic determinism and, 63–64; hidden variables and, 54–57; increased acceptance of, 49; locality and, 71–72; local realism and, 73–74; Maharishi Mashesh Yogi and, 83–84; old vs. new, 30–32; origins of, 4–5; parapsychology and, 131; "participator" aspects of, 130–31; philosophical and practical applications, 51; popular culture and, vii–viii, 1, 5–7, 77–81; probability amplitudes, 116–18; "reality" of, 45–46; spin and angular momen-

quantum theory (*cont.*): tum, 57–62; statistical interpretation (wave function), 38; "umpire" analogy of (Wheeler), 96; wave and matrix mechanics and, 34; Wheeler's view of, 96, 131–34. *See also* solid-state (condensed-matter) theory
Queens University, 13–14
"The Queerness of Quanta," 5

Rabi, I. I., vii, 60, 93–94, 104
radioactivity, spin concept and, 61
Rayleigh-Jeans law, 23, 26
reality and quantum theory: coin toss analogy, 45–46; EPR experiment, 45–46; philosophical aspects of physics and, 94–95; quantum theory and, 42–43, 84–85, 140–41; spin concept and, 61. *See also* local reality
relativity theory: Einstein-Besso discussions of, 147, 152–53; Einstein's 1905 paper on, 4, 143; gravitation and, 43–44; planetary atomic model and, 31–32; role of time in, 161; speed of light and, 70–71; spin concept and, 58–59; Wheeler's view of, 133–34
religion, physics and, 53. *See also* Eastern religion
Reviews of Modern Physics, 112
Rockefeller Institute for Medical Research, 127
Roemer, Ole, 70
Rosenfeld, Léon, 43, 111; EPR experiment, 46–47
Rosen, Nathan, 44–45
Royal Prussian Academy of Sciences, 27
Ruark, Arthur, 109
Rutherford, Ernest, 28–29, 104, 107

sabotage and atomic research, 125–26
safety, of nuclear plants, 119–22, 124–25
Schrödinger equation, 33–37, 95, 161
Schrödinger, Erwin: Bohr debates with, 55–56; cat paradox of quantum mechanics, 50–51; education and personal life, 32–33; Einstein nominates to Nobel Prize, 44; Einstein's relations with, 146, 160; interest in Eastern religions, 80; opposition to statistical mechanics, 38; quantum theory and, 55; origins of quantum theory and, 4
semi-empirical mass formula, 111–12
Seven Years in Tibet, 80
Shimony, Abner, 75–76
Simon, Walt, 122–23
Skitt, David, 81–82
slow-neutron fission, 112–13
Snyder, Hartland, 16, 117–18, 129
solid-state (condensed-matter) theory, 27–28
Solovine, Maurice, 3
Solvay Congress, 43
Solvay, Ernest, 40
Sommerfeld, Arnold, 143
space quantization, 60
spacetime concept, 133–34
Speakable and Unspeakable in Quantum Mechanics, 8
Speziali, Pierre, 147–49, 154
"spin" concept, 57–62; Bell's inequality and, 75; values in, 59
Spitzer, Lyman, 127
"spooky actions at a distance," 44, 62, 73
"Stability of Perturbed Orbits in the Synchrotron," 16–17
Stanford Linear Accelerator, 67–68
statistical mechanics, 26, 140, 151
stellarator research, 127
Stern, Otto, 60–61, 76, 160
Stoppard, Tom, 5
Strassman, Fritz, 110–11

strong focusing principle, 16–17
Super Collider project, 11
"superluminal" phenomena, 78
superspace concept, 133–34
Surely You're Joking, Mr. Feynman!, 97–98
Surprises in Theoretical Physics, 67
Swiss Federal Patent Office, 147–48, 152–53
synchrotron, defined, 16–17
Szilard, Leo, 113

tachyons, 72
TCP theorem, 17–18
telepathy, quantum theory and, 79
Teller, Edward, 127
thermodynamics, 22; hidden variables and, 54, 64
"The Science of Mechanics," 145
Thompson, G. P., 14, 31
Thomson, J. J., 14, 28–29
Thorne, Kip, 140–41
three-dimensional geometries, 133–34
Toll, John, 128
Tucci, Niccolò, 150
Turner, Louis, 112
twin analogy of correlations, 62–63

Ufford, Letitia Wheeler, 109
Uhlenbeck, George, 58–59
Ulam, Stanislaw, 127
umpire analogy and quantum theory (Wheeler), 96, 117, 139–40
uncertainty principle (Heisenberg): Bell puzzled by, 50–51; determinism and, 35–37; double-slit experiment, 41–42; energy-time aspects, 43–44; popular culture and, 5; quantum theory and, 140–41; spin concept and, 59, 61–62; Z^0 principle, 90–91
Unified field theory, 110, 157–58, 160
University of North Carolina, 109
University of Texas, 129–30
University of Wisconsin, 110
University of Zurich, 153
Upanishads, 6
uranium isotopes, and slow-neutron fission, 112

von Neumann, John, 56; theorem, 64–65, 67

wave equation, superspace and, 134
wave function, 35; "state" of electrons and, 95
wave mechanics, 33–34
wave theory of light, 24–25; principle of complementarity and, 42
Weyl, Hermann, 104, 137
What Do You Care What Other People Think?, 116
What is Life?, 33
Wheeler, John Archibald: atomic bomb research, 113–14, 117–18, 124; black hole research, 129–30; Bohr's influence on, 107–8, 126; bypass surgery on, 135–36; childhood of, 100–101; constructs fallout shelter, 128–29; educational career, 101–4; Einstein assessed by, 138–41; family life and history, 98–100, 109–10; Feynman and, 97–98, 107, 114–16; Fuchs and, 125–26; geon experiments, 132–33; Guggenheim Fellowship awarded to, 126; at Hanford Engineering Works, 119–24; hydrogen bomb research and, 126–29; interest in explosives, 102; marriage to Janette Hagner, 107; nuclear fission research, 110–18; nuclear reactor research, 118–24; Ph.D. thesis of, 104–6; philosophical aspects of physics, 94, 130–31; Princeton University post, 109–10, 124–30; quantum theory interpretation, 9, 95–96, 131–34, 136–41; receives

Wheeler, John Archibald (*cont.*)
National Research Council Fellowship, 105–7; research assessed, 98; retirement, 130, 135–36; social life, 106–7; teaching career assessed, 93–98, 130, 136–37; at University of Texas, 129–30
"Wheeler-Moke Safe and Gun Company," 101
Wien, Wilhelm, 22, 26
Wigner, Eugene, 109–10, 113, 137–38
Wilczek, Frank, 129
Williams, E. J., 107–8
Williams, Roger, 119–22
Winteler, Barbara, 149
Winteler, Paul, 149–50

Young, Thomas, 24–25, 30, 40–41

Zeitschrift für Physik, 103
Zukav, Gary, 6, 77–78, 130–31
Zurek, Wojciech, 9
Z^0 width concept, 90–91